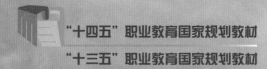

"十四五"职业教育国家规划教材

"十三五"职业教育国家规划教材

U0201735

飞机维修专业英语

——飞机系统 第二版

主　编◇赵迎春

副主编◇魏　敏　陈凯军　郭　俊

主　审◇杨国余

中国水利水电出版社

www.waterpub.com.cn

·北京·

内 容 提 要

本书由两大模块组成：模块一为机械系统，通过 7 个学习情境对空调与增压系统、飞机操作系统、燃油系统、液压系统、气动系统、饮用水及废水系统和发动机润滑油进行了专业的英文介绍；模块二为航电系统，通过 7 个学习情境对飞机上的通信系统、电源系统、外部电源、防火、灯光系统、导航系统、警告系统进行了专业的英文介绍。这些英语学习情境结合航空专业材料，通过听、说、读、写、译的语言训练，使学生轻松掌握飞机主要系统的英文术语、专业词汇、缩略词、高频句型与难句，并读懂英文资料。

本书既可以作为航空类本科与高职院校飞机维修专业、发动机维修专业及相关专业学生的专业英语教材，也可以作为航空公司机务工程部及飞机维修公司在职人员的专业英语培训和自学教材。

图书在版编目（CIP）数据

飞机维修专业英语. 飞机系统 / 赵迎春主编. -- 2版. -- 北京：中国水利水电出版社，2022.5（2024.8 重印）
"十三五"职业教育国家规划教材
ISBN 978-7-5226-0618-7

Ⅰ．①飞… Ⅱ．①赵… Ⅲ．①飞机－维修－英语－高等职业教育－教材 Ⅳ．①V267

中国版本图书馆CIP数据核字(2022)第063415号

策划编辑：周益丹　责任编辑：鞠向超　加工编辑：刘　瑜　封面设计：梁　燕

书　名	"十三五"职业教育国家规划教材 "十四五"职业教育国家规划教材 **飞机维修专业英语——飞机系统（第二版）** FEIJI WEIXIU ZHUANYE YINGYU——FEIJI XITONG
作　者	主　编　赵迎春 副主编　魏　敏　陈凯军　郭　俊 主　审　杨国余
出版发行	中国水利水电出版社 （北京市海淀区玉渊潭南路 1 号 D 座　100038） 网址：www.waterpub.com.cn E-mail：mchannel@263.net（答疑） 　　　　sales@mwr.gov.cn 电话：（010）68545888（营销中心）、82562819（组稿）
经　售	北京科水图书销售有限公司 电话：（010）68545874、63202643 全国各地新华书店和相关出版物销售网点
排　版	北京万水电子信息有限公司
印　刷	雅迪云印（天津）科技有限公司
规　格	184mm×260mm　16 开本　17.5 印张　480 千字
版　次	2019 年 2 月第 1 版　2019 年 2 月第 1 次印刷 2022 年 5 月第 2 版　2024 年 8 月第 6 次印刷
印　数	13001—18000 册
定　价	62.00 元

第二版前言

当前，我国航空事业迅速崛起，中国共产党二十大报告充分肯定国产大飞机制造取得的重大成果。C919 飞机的问世及国外确认订单的迅速增长，这个事件意义重大，为我国航空事业走向国际增添了信心与力量。在航空维修领域，培养与世界接轨的国际化高素质技术技能型机务人才，是我国航空事业国际化发展的重要保障，为中国航空大国走向航空强国提供重要支撑。正是基于航空产业高度国际化特征，众多航空公司、机场及飞机维修基地对机务人员在航空维修领域的英语阅读和交际能力提出了更高要求，以适应航空维修高质量发展需求，机务人才国际化水平也是其保障飞机安全，实现人生价值，达成"航空报国"理想的利刃。

作为我国机务人才培养基地的各类航空类职业院校，已经充分意识到航空维修专业英语素质的重要性，这些职业院校加大了航空维修专业英语教材建设与开发，在教材编写理念、整体设计、素材选取、内容编排等方面取得了明显进步，尤其改变了以阅读能力为主的传统观点，体现了岗课对接的基本思想，但总体而言，依然在岗位针对性、语言能力培养系统性、内容时效性、思政元素的有机融合等方面依然存在不足。

本教材于 2019 年 2 月第一次出版，在三年实践过程中，编写组广泛听取相关院校教师、航空企业专家以及毕业生的意见和建议，第二版保留了原教材设计的基本理念、整体框架及核心内容的基础上进行了较大修订与完善，修改如下：

（1）融入中国民航局最新标准。根据 2020 年 6 月中国民航局飞行标准司公布的《航空维修技术英语等级测试指南》（AC-66-FS-010），根据考核点及题型重新设计课后练习，同时同步最新技术，更换了部分机型实物图及原理图。

（2）新增警示案例及二维码清单。每课增加了航空警示案例；新增的二维码清单清晰明了，内含大量微课、文本资源，融入了航空强国、航空报国、三敬精神、航修金牌工匠等内容。

（3）新增机务场景对话。为服务国家"一带一路"倡议，满足中国航空企业高质量发展"走出去"要求，"航空听说"在原内容上增加了航空修理场景对话。

（4）新增"航空应用"。在原"航空听说""航空阅读""航空翻译""航空写作"的基础上，增加了"航空应用"——《飞机维修手册》实操工卡阅读，并于每课后随附"学习反思"。

（5）配套中国大学 MOOC"飞行器维修专业英语"线上课程，进一步打造线上线下混合教学模式。

本教材经过修订改版后，主要体现的特色如下：

1. 基于飞机维修岗位能力需求，开发学习模块，动态更新行业最新标准

本教材学习情境参考了机务维修岗位对英语的具体要求，融入了原版技术文献内容，按照 CCAR-66R3 部、CCAR-147-02 部规章相关咨询通告的考试及培训大纲要求，旨在让学生掌握飞机系统的英文术语、专业词汇、缩略词、高频句型与难句，能读懂机务维修岗位上的原版技术文献，完成必要的口头及书面交际。在内容及形式上与时俱进，融入中国民航局最新标准及航修岗位需求，课后练习的题型及考点均根据 2020 年 6 月中国民航局飞行标准司公布的《航空维修技术英语等级测试指南》（AC-66-FS-010）进行设计。

2. 基于 ATA-100 国际标准，编排飞机系统学习情境

学习情境按照岗位的具体工作先分为航空机械和航空电气电子两个部分，再按照 ATA-100（美国航空运输协会 100 号规范）出现的先后顺序依次安排。目的是既让学生熟悉了该部分的英文术语、缩略语、手册中出现的各类大写警示语，同时又熟悉了飞机主要系统、ATA-100 的具体内容，并在教材中增加实物图及原理图，图文并茂，使系统学习更直观、高效。

3. 按照英语"听、说、读、写、译"五大能力，无缝对接岗位英语实际需求

学习情境由飞机某系统的"航空听说""航空阅读""航空翻译""航空写作""航空应用"五个部分组成。"航空听说"，立足中国，沟通世界，凸显专业英语交际能力培养，为响应党二十大精神，服务国家"一带一路"倡议，满足中国航空企业高质量发展"走出去"要求，将该部分分为航空听力及航空修理场景对话两部分。"航空阅读"旨在让学生熟悉飞机系统中的英文术语、高频词汇、缩略语、高频句型与难句，使学生具备读懂全英文文献的能力。"航空翻译"旨在让学生读懂英文手册中大写警示句，提高维修工作中的安全及规范操作意识；"航空写作"是为了让学生在维修工作中能够用英文进行维修故障描述、故障诊断并提出维修建议，"航空应用"立足基础，强化应用，进一步提升岗位针对性，材料节选自《飞机维修手册》各种实操程序，根据在岗机务维修人员填报英文工卡的实际需求设计，检验学生活学活用的能力。

4. 显隐同行，双线并重，强化课程思政教育

优秀的专业英语教材能让学生阅读外国技术文献，完成工作中的正常交际，这是英语的工具性。而当异域文化与本土文化发生冲突，能引导学生自动践行社会主义核心价值观，树立正确的人生观，体现文化自信，这是英语的人文性。为了完美融合工具性及人文性，本教材及配套的立体化资源的思政教育以语言线及职业线双线并展，合二为一，主线融合了坚守奉献、开拓创新的职业理念，以"用英语讲好中国航修故事，培育航修工匠情怀"为目标，语言线在培养其职场涉外沟通、多元文化交流、语言思维提升及自主学习完善能力的同时，使其践行社会主义核心价值观。职业线通过微课、文本等形式融入航修政策、航修理想、航修人物、航修故事、警示案例、航修新闻、航空装备（中国制造）的民族自豪感等，提高航修工作中的安全、责任、遵守规范意识，注重培养学生爱岗敬业、航修报国的职业态度及理想，让其逐渐养成"重安全、零缺陷"职业思想，秉承"仔细阅读英文、悟透操作规范"的"细致、认真、负责"的工匠精神，培养其作为未来航修人的责任感、使命感及创新精神和卓越精神。

5. 线上线下，无缝对接，进一步打造线上线下混合教学模式

配套立体化资源，拓展全方位学习渠道。从资源形式来看，有录音、微课、文本，从空间来看，系统构建了以"纸质教材"为中心，以"教材内嵌二维码资源＋在线课程＋微信公众号"三层立体化教学资源。录音包含了听力、对话、单词及阅读材料录音，微课包含专业知识点、词汇、句型、职业素养系列微课。专业知识点微课包含手册学习及飞机系统讲解，词汇系列包含构词法专讲及术语专讲，句型系列包含长难句翻译方法及难句分析。文本资源包含练习答案、译文、警示案例等。

线上资源丰富，与纸质教材互为补充，满足学生自主学习及个性化学习需要。教材内嵌资源二维码，扫码可得，在线课程"飞行器维修专业英语"在中国大学 MOOC 搜索可得，同时，搜索微信公众号"航空维修专业英语"，可得部分更新资源，另外本书配有"航空维修专业英语"题库，授课教师如有任何疑问或资源需求，可发送邮件至 617270095@qq.com 进行交流。

本教材由赵迎春副教授担任主编，中国南方航空公司机务培训中心质量经理、全国发动机维修技能大赛裁判长杨国余先生担任主审，魏敏、陈凯军、郭俊老师担任副主编，CCAR147 机务培训中心质量经理江游、皮芳、刘逸众、王爱娥老师参与了本书编写。长沙航空职业技术学院主管教学副院长朱国军、教务处处长熊纯、基础教育学院院长唐倩、机电学院院长陈律对本书给予了大量支持及帮助。在此，对为本书编写给予指导、支持、帮助的领导、同事、学生及所有提供英文技术资料的机务维修一线的领导和技术人员表示诚挚的感谢，特别感谢中国民航大学张文林老师的飞机维护虚拟仿真训练系统为本教材提供宝贵资源。

尽管编者本着"有错必纠"的原则，本书还是难免存在疏漏之处，衷心感谢江苏航空职业技术学院、日照职业技术学院等兄弟院校为本书纠错，也恳请读者、同行批评指正，以便于本书在今后修订过程中的改进。

扫码看中国大学 MOOC
飞行器维修技术专业英语

编者

2022 年 2 月

第一版前言

近年来，随着我国民航事业的迅猛发展与民航维修标准的国际化，业内对飞机维修人员的英语阅读、翻译水平和交际能力提出了更高的要求。而随着航空产业的大力发展，目前众多航空公司、机场及飞机维修基地的在职机务人员的专业英语综合能力诸如口头交际和书面沟通能力有待提高。

目前，我国机务人才培养基地在各级航空类院校，这些学院已经认识到一本优秀的专业英语教材对提升学生专业英语综合能力的重要性。遗憾的是，目前市场上与岗位零对接且体现英语综合能力的专业英语教材不多，大部分教材偏重阅读，阅读材料针对岗位性不强，缺乏听、说、写、译能力的系统训练，不能满足岗位对英语综合能力的要求。部分院校节选《飞机维修手册》部分章节的复印资料作为教材，尽管针对性强，但缺乏系统性、全面性，又缺乏理论基础，更无配套练习；同时，大部分教材注重文字表述，缺少实物插图，而航空类专业英语词汇与术语非常抽象，涉及机械、通讯、电子、电源、液压等诸多专业领域，没有配图的辅助，学生很难记住这些英语词汇。

本教材根据机务工作的实际需求和工作情境，针对飞机维修及相关专业对英语的个性化需要，参照中国民航局颁发的《CCAR-147-02部民用航空维修基础培训大纲》《CCAR-66部维修人员培训大纲》及各航空公司相关岗位的实际需求，借鉴《飞机维修手册》部分内容，再结合其他英文资料，配以实物插图，编写了集听、说、读、写、译能力为一体的《飞机维修专业英语——飞机系统》。本教材继承了传统专业英语教材注重阅读与翻译能力培养的特点，同时加大了对航空专业领域内英语听力、口语表达及英文写作的培养力度，具有如下鲜明的特色：

1. 基于飞机维修职业能力需求，开发学习模块

本教材参考了机务人员岗位对英语的具体要求，按照CCAR-66部、CCAR-147部规章相关咨询通告的考试和培训大纲要求，以熟悉飞机主要系统的英文术语、专业词汇、缩略词、高频句型与难句，达到让学生能读懂关于飞机系统的英文资料为目的，开发了飞机主要系统的学习模块。

2. 基于ATA-100，编排飞机系统学习情境

本模块的学习情境按照岗位的具体工作先分为航空机械和航空电气电子两个部分，再按照ATA100中出现的先后顺序依次安排。这样安排既让学生熟悉了该部分的英文专业术语、手册中出现的诸如各类警告的大写英文材料、各种缩略语，同时又熟悉了飞机上的主要系统、ATA100的具体内容，体现了英语作为工具的重要性。

3. 按照英语"听、说、读、写、译"五大能力，细化专业学习内容

每一个学习情境由飞机的某一个系统的英语"听说""阅读""翻译""工卡写作"四

大部分组成。"听说"为能更好地直接用英文交流而设计；"阅读"与"翻译"主要让学生熟悉飞机系统中的各种英文术语、高频词汇、缩略语、高频句型和难句，熟悉工作中的各种英文文献，使学生具备读懂全英文文献的能力；"工卡写作"是为了让学生在维修工作中能够用英文做出维修故障描述，维修故障诊断并提出维修故障建议而设计。

4. 无缝对接岗位需求与实际学习，体现"工学结合"

本教材选用的飞机的主要系统，是维修工作中需要首先了解的系统，所选的材料大部分来自于实际工作中的维修文献，所有练习的设计是为了让学生能够快速熟悉飞机系统的英文专业词汇、术语、缩略语和句型，相关视频也是来自于国际和国内一线工作的全英文视频。

5. 抽象的系统结合实物及原理图解，图文并茂

本教材的阅读文章通过专业讲解与实物插图的结合，图文并茂地将飞机主要系统展现出来，生动形象，使得学生能轻易把抽象的系统学习与原理图相联系，使系统的学习更加直观和高效。

6. 开发立体化教材，拓展全方位学习渠道

本教材为学生提供了多渠道、多元化的学习资料，涵盖了电子教案、PPT、闯关卡、微课、在线上课等各种资料。教材的内容、答案、录音原文、录音都可通过扫码获取，使用方便，打破了以往传统英语教材形式的单一性，满足自主学习的要求。

本教材可作为本科类和高职类航空院校的发动机维修、飞机维修等专业的专业英语教材使用；也可作为航空电子设备维修、通用航空器维修、无人机飞行器、理化测试与质检等专业的向民航拓展的专业英语资料；亦可作为所有民航在职机务人员、致力于向民航拓展的航修人员的专业英文自学材料。本教材可与《飞机维修专业英语教程——飞机主要部件》作为系列教材使用，亦可单独使用。

本教材由赵迎春副教授、陈凯军副教授担任主编，中国南方航空公司机务培训中心质量经理、全国发动机维修技能大赛裁判长杨国余先生担任主审，魏敏、王志敏、王江、罗娜老师担任副主编，姜维、曾娅妮、解芳、唐启东、刘逸众、李敏老师参与了教材编写。主管教学副院长朱国军、基础教育学院院长唐倩、机电学院院长熊纯、CCAR147机务培训中心质量经理江游、教务处处长彭圣文对本教材给予了大量技术性支持和帮助。在此，对本教材编写给予指导、支持、帮助的领导、专业英语教研室同事、飞机教研室同事，以及所有提供过帮助的学生表示真诚的感谢。在编写的过程中，本书参考了大量的书籍、资料和各种型号的飞机维修手册，在这里谨向本书提供帮助的机务维修一线的领导和技术人员表示诚挚的感谢。

由于时间仓促，编者实践经验有限，本教材难免存在错误，编者会在本书的使用过程中仔细勘查，并一一校正。

编者

2019 年 7 月

目　录

第二版前言
第一版前言

Module One　Mechanic Systems

Lesson 1 Air Conditioning and Pressurization System..2

Learning Objectives 2

Section 1 Aviation Listening & Speaking 2

Section 2 Aviation Reading..4

Section 3 Aviation Translation: translate the following sentences into Chinese................................14

Section 4 Aviation Writing..14

Lesson 2 Flight Control System....................17

Learning Objectives 17

Section 1 Aviation Listening & Speaking 17

Section 2 Aviation Reading..19

Section 3 Aviation Translation: translate the following sentences into Chinese................................28

Section 4 Aviation Writing..28

Lesson 3 Fuel System..................................31

Learning Objectives 31

Section 1 Aviation Listening & Speaking 31

Section 2 Aviation Reading..33

Section 3 Aviation Translation: translate the following sentences into Chinese................................43

Section 4 Aviation Writing..43

Lesson 4 Hydraulic System.........................46

Learning Objectives 46

Section 1 Aviation Listening & Speaking 46

Section 2 Aviation Reading..48

Section 3 Aviation Translation: translate the following sentences into Chinese................................58

Section 4 Aviation Writing..58

Lesson 5 Pneumatic System........................61

Learning Objectives 61

Section 1 Aviation Listening & Speaking 61

Section 2 Aviation Reading..62

Section 3 Aviation Translation: translate the following sentences into Chinese................................72

Section 4 Aviation Writing..72

Lesson 6 Water and Waste System 74

Learning Objectives 74

Section 1 Aviation Listening & Speaking 74

Section 2 Aviation Reading..75

Section 3 Aviation Translation: translate the following sentences into Chinese................................84

Section 4 Aviation Writing..85

Lesson 7 Engine Oil 87

Learning Objectives 87

Section 1 Aviation Listening & Speaking 87

Section 2 Aviation Reading..89

Section 3 Aviation Translation: translate the following sentences into Chinese................................97

Section 4 Aviation Writing..98

Module Two   Electrical and Avionic Systems

Lesson 8   Communication System 101
Learning Objectives 101
Section 1 Aviation Listening & Speaking 101
Section 2 Aviation Reading 103
Section 3 Aviation Translation: translate the following
          sentences into Chinese 118
Section 4 Aviation Writing 118

Lesson 9   Electrical Power System 121
Learning Objectives 121
Section 1 Aviation Listening & Speaking 121
Section 2 Aviation Reading 123
Section 3 Aviation Translation: translate the following
          sentences into Chinese 138
Section 4 Aviation Writing 138

Lesson 10   External Power 140
Learning Objectives 140
Section 1 Aviation Listening & Speaking 140
Section 2 Aviation Reading 141
Section 3 Aviation Translation: translate the following
          sentences into Chinese 149
Section 4 Aviation Writing 149

Lesson 11   Fire Protection 151
Learning Objectives 151
Section 1 Aviation Listening & Speaking 151
Section 2 Aviation Reading 153
Section 3 Aviation Translation: translate the following
          sentences into Chinese 167
Section 4 Aviation Writing 167

Lesson 12   Lighting System 169
Learning Objectives 169
Section 1 Aviation Listening & Speaking 169
Section 2 Aviation Reading 171
Section 3 Aviation Translation: translate the following
          sentences into Chinese 182
Section 4 Aviation Writing 182

Lesson 13   Navigation System 185
Learning Objectives 185
Section 1 Aviation Listening & Speaking 185
Section 2 Aviation Reading 187
Section 3 Aviation Translation: translate the following
          sentences into Chinese 198
Section 4 Aviation Writing 198

Lesson 14   Warning System 201
Learning Objectives 201
Section 1 Aviation Listening & Speaking 201
Section 2 Aviation Reading 202
Section 3 Aviation Translation: translate the following
          sentences in Chinese 211
Section 4 Aviation Writing 211

Appendix I   Frequently-used Tools in Chinese &
            English 213
Appendix II   Organizations & Related Terms in
            Aircraft M & O 222
Appendix III Examples of AMM-PP 224
Appendix IV My Leaflet Workbook 253
References 267

QR Code List（二维码清单）

Lesson 1 Air Conditioning and Pressurization
System..2

1 听力录音 2

2 答案及原文 2

3 对话录音 3

4 对话译文 3

5 课文朗读 4

6 课文译文 4

7 空调系统（微课）...................................5

8 单词朗读 11

9 航空术语学习策略（微课）...................11

10 答案 12

11 三敬精神——感动中国 2018 年度人物
刘传健（文本）...................................14

12 警示案例——川航 3U8633- 生与死 16

13 PP 译文 224

14 中国民航局《航空维修技术英语》（微课）........224

Lesson 2 Flight Control System.................17

1 听力录音 17

2 答案及原文 17

3 对话录音 18

4 对话译文 18

5 课文朗读 19

6 课文译文 19

7 飞行操纵系统 (微课)...........................19

8 单词朗读 25

9 术语专讲：蹬舵（微课）.....................25

10 答案 26

11 岗位自豪（文本）.............................28

12 警示案例——狮子航空 610 雏鹰折翅 30

13 PP 译文 226

14 怎样获取 FAA 飞机维修执照（微课）........226

Lesson 3 Fuel System.............................31

1 听力录音 31

2 答案及原文 31

3 对话录音 32

4 对话译文 32

5 课文朗读 33

6 课文译文 33

7 发动机燃油（微课）.............................33

8 单词朗读 40

9 术语专讲：空中加油（微课）...............40

10 答案 41

11 精致维修——机务团队的"工匠精神"（文本）..43

12 警示案例——沙佩科人空难：燃油不足 45

13 PP 译文 228

14 机务常用英文技术资料（微课）...........228

Lesson 4 Hydraulic System.........................46

1 听力录音 46

2 答案及原文 46

3 对话录音 47

4 对话译文 47

5 课文朗读 48

6 课文译文 48

7 飞机上的液压系统（微课）...................48

8 单词朗读 55

9 术语专讲：液压（微课）.....................55

10 答案 56

11 空军"金牌蓝天工匠"施娟：成功的人
舍得真心付出 58

12 警示案例——联合航空 232 号班机 苏城空难........60

13 PP 译文 230

14 ATA-100 介绍（微课）.........................230

Lesson 5 Pneumatic System.......................61

1 听力录音 61

2 答案及原文 61

3 对话录音 62

4 对话译文 62

5 课文朗读 62

6 课文译文 62

7 气动系统（微课） 63

8 单词朗读 69

9 术语专讲：舱（微课） 69

10 答案 70

11 忘填工卡日志（文本） 72

12 警示案例——MU5759 排气系统故障 73

13 PP 译文 232

14 飞机维修手册（微课） 232

Lesson 6 Water and Waste System 74

1 听力录音 74

2 答案及原文 74

3 对话录音 75

4 对话译文 75

5 课文朗读 75

6 课文译文 75

7 水 & 废水系统（微课） 76

8 单词朗读 82

9 术语专讲：真空（微课） 82

10 答案 83

11 航修焊接专家孙红梅—— 一颗匠心护 "战鹰"
（文本） 84

12 警示案例——加拿大航空 86

13 PP 译文 234

14 飞机维修手册 SDS（微课） 234

Lesson 7 Engine Oil 87

1 听力录音 87

2 答案及原文 87

3 对话录音 88

4 对话译文 88

5 课文朗读 89

6 课文译文 89

7 发动机滑油（微课） 89

8 单词朗读 95

9 术语专讲：火花塞（微课） 95

10 答案 96

11 航修手艺人（文本） 97

12 警示案例——阿拉斯加航空 261 号班机—无润滑油
的螺纹磨秃 99

13 PP 译文 236

14 飞机维修手册——PP（微课） 236

Lesson 8 Communication System 101

1 听力录音 101

2 答案及原文 101

3 对话录音 102

4 对话译文 102

5 课文朗读 103

6 课文译文 103

7 飞机的通信系统（微课） 103

8 单词朗读 115

9 术语专讲：雷达（微课） 115

10 答案 116

11 航空安全警示语（文本） 118

12 警示案例——特内里费空难 120

13 PP 译文 238

14 英文手册的警告—告诫—注意辨析（微课） 238

Lesson 9 Electrical Power System 121

1 听力录音 121

2 答案及原文 121

3 对话录音 122

4 对话译文 122

5 课文朗读 123

6 课文译文 123

7 飞机上的电源系统（微课） 123

8 单词朗读 135

9 术语专讲：继电器（微课） 135

10 答案 136

11 中国梦·大国工匠篇——航修专家罗卓红：
要干就干得最好 138

12 警示案例——美国大陆快运 2574 号班机 ... 139

13 PP 译文 240

14 页块（微课） 240

Lesson 10 External Power.........................140

1 听力录音 140

2 答案及原文 140

3 对话录音 141

4 对话译文 141

5 课文朗读 141

6 课文译文 141

7 外部电源（微课） 142

8 单词朗读 147

9 术语专讲：计量表（微课） 147

10 答案 147

11 坚守岗位，不负使命——海航集团金鹏航空维修
技术员罗湘军用脚步践行（文本）......... 149

12 警示案例——卓克斯海洋航空 101 号班机—
右翼断裂 150

13 PP 译文 242

14 计划性维护与非计划性维护文献（微课）......... 242

Lesson 11　Fire Protection 151

1 听力录音 151

2 答案及原文 151

3 对话录音 152

4 对话译文 152

5 课文朗读 153

6 课文译文 153

7 防火（微课）.............................. 153

8 单词朗读 163

9 术语专讲：管道（微课）.................... 163

10 答案 164

11 中国制造——鲲龙（文本）................. 167

12 警示案例——瑞士航空 111—机舱烈焰 168

13 PP 译文 244

14《故障隔离手册》（微课）................. 244

Lesson 12　Lighting System 169

1 听力录音 169

2 答案及原文 169

3 对话录音 170

4 对话译文 170

5 课文朗读 171

6 课文译文 171

7 外部灯光系统 & 灯光子系统（微课）......... 171

8 单词朗读 178

9 术语专讲：电门（微课）.................... 178

10 答案 180

11 中国制造——C919（文本）................ 182

12 警示案例——东航 401—致命灯泡 184

13 PP 译文 247

14 飞机的种类（微课）...................... 247

Lesson 13　Navigation System 185

1 听力录音 185

2 答案及原文 185

3 对话录音 186

4 对话译文 186

5 课文朗读 187

6 课文译文 187

7 飞机上的导航系统（微课）................. 187

8 单词朗读 195

9 术语专讲：雷达波瓣（微课）............... 195

10 答案 196

11 情注蓝天　航修报国—李天（文本）......... 198

12 警示案例——大韩航空货运 8509—刚愎自用....... 200

13 PP 译文 249

14 波音 737 经典的一代 VS 次世代（微课）........... 249

Lesson 14　Warning System 201

1 听力录音 201

2 答案及原文 201

3 对话录音 202

4 对话译文 202

5 课文朗读 202

6 课文译文 202

7 飞机上的警告系统（微课）................. 203

8 单词朗读 208

9 术语专讲：复飞（微课）.................... 208

10 答案 209

11 一名航修"老兵"的依依深情（文本）................. 211

12 警示案例——乌柏林根空难 212

13 PP 译文 250

14 旋翼飞机主要结构（微课）................. 250

Module One   Mechanic Systems

Introduction

Aircraft systems are complex. In the design stage and in the operating process to ensure continued airworthiness of an aircraft, they are broken down into simpler subsystems that carry out homogeneous functions. Some examples include but are not limited to:

- Pressurization system. Flight controls.
- Hydraulics. Pneumatic system. Electrical system. Bleed system.
- Water and waste system. Supplemental oxygen. Avionics.
- Navigation.
- Communication.
- Ice protection (anti-icing and de-icing).
- Instrumentation and recording.
- Fire protection. Safety system.

This module covers the essential mechanic and avionic knowledge base required by certifying mechanics, technicians and engineers engaged in engineering maintenance activities on commercial aircraft and in general aviation.

Lesson 1　Air Conditioning and Pressurization System

Learning Objectives

1. Knowledge objectives:

    A. To grasp the words, related terms and abbreviations about air conditioning & pressurization system.

    B. To grasp the key sentences about air conditioning & pressurization system.

    C. To know the main components and functions of air conditioning & pressurization system.

2. Competence objectives:

    A. To be able to read and understand frequently-used & complex sentence patterns, capitalized English materials and obtain key information quickly.

    B. To be able to communicate with English speakers about the topic freely.

    C. To be able to fill in the job cards in English.

3. Quality objectives:

    A. To be able to self-study with the help of aviation dictionaries, the Internet and other resources.

    B. To develop the craftsman spirit of carefulness and responsibility.

Section 1　Aviation Listening & Speaking

1.1　Aviation Listening: listen to the record and fill in the blanks.

1 听力录音　　2 答案及原文

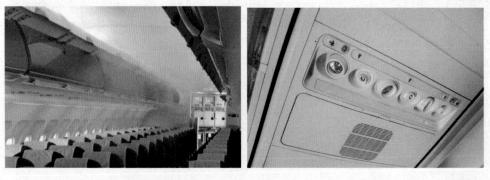

    In this lesson, we will look at the 1._____of air for an aircraft cabin-air 2._____system. And we will examine the methods used to maintain the air at a 3._____temperature. We will begin, however, with a short review of the makeup of the atmosphere, to explain the need for air conditioning and 4._____systems. Modern jet airliners cruise at high 5._____up to 43,000 feet. The advantages of high altitude flying are greater fuel efficiency and higher speed. And on most occasions, the aircraft will be above any bad weather. The disadvantages are that there is 6._____

    7._____to breathe and the air is extremely cold in a standard atmosphere, the temperature drops at a rate of 1.98 degrees per thousand feet up to the tropopause, which is at approximately

36,000 feet or eleven kilometers. This means that on a standard day, at 31,000 feet, the outside air temperature will be a 8._____ minus forty-six degrees Celsius. The atmosphere is made up of 78% 9._____and 21% oxygen, with the other 1% consisting of various trace 10._____.

1.2 Aviation Speaking: practice the following dialogue and design your own.

Situation: This is a dialogue between the pilot and the ground mechanic. After starting, the pilot calls the ground that there was a fault in the cockpit. And then the plane is ready to be towed to a set position. Practice it and have your own.

Note: PIL=Pilot, GND=Ground Mechanic

3 对话录音

Scene 1

PIL: Ground, Cockpit.

GND: Go ahead.

PIL: The left PACK light illuminates when recalling.

GND: Roger. Standby. I need to report the fault to the engineer now.

4 对话译文

Scene 2

PIL: Cockpit to ground.

GND: Go ahead.

PIL: Please call a tractor.

GND: Roger, call a tractor.

PIL: Please guide us to a parking place.

GND: Copy that, to a parking place.

PIL: Please connect the tow-bar.

GND: Roger, tow-bar set. Please set the parking brake.

PIL: OK, parking brake is set. Remove the chock.

GND: Chock is clear. Release the parking brake.

PIL: Parking brake released. Please make sure that the towing speed is not above 5 kilometers per hour.

GND: OK, no problem.

PIL: Clear for towing, it is not into position, a little more forward please.

GND: Got it, a little forward.

Words & Expressions

1. tow [təʊ] *v.*   牵引
2. illuminate [ɪˈluːmɪneɪt] *v.*   点亮，照亮
3. go ahead   请讲
4. tractor [ˈtræktə(r)] *n.*   牵引车
5. copy that   收到

6. release [rɪˈliːs] v.　　　　松开
7. clear for towing　　　　　允许牵引

Section 2　Aviation Reading

5 课文朗读　　6 课文译文

Pre-reading questions:

1. What do you know about the air conditioning system?
2. Do you know the functions of pressurization system?

Air Conditioning and Pressurization System

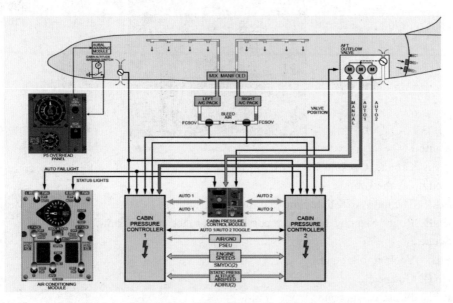

Fig.1-1　Schematic for B737-700 Air Conditioning System

Introduction

The air conditioning system supplies conditioned air for heating and cooling the cockpit and cabin spaces for the passengers and crew. It makes sure that the air is fresh and clean, at a comfortable temperature, the correct humidity and the correct pressure. It also heats the cargo compartment and cools the electronic equipment. This air also provides pressurization to maintain a safe and comfortable cabin environment. In addition to cabin air conditioning, some aircraft equipments and equipment compartments require air conditioning to prevent heat buildup and consequent damage to the equipment.

The air conditioning packs supply conditioned air into pressurization system. This system also supplies pressurization to the whole passenger compartment, the cockpit, all the cargo compartments and the EE compartment. Landing-gear bays, the radome and the tail cone are not pressurized and they are separated by pressure bulkheads.

The pressurization system controls cabin altitude by regulating air flow from the fuselage through the outflow valve. The pressurization system maintains cabin altitude as close to sea level as possible or at an altitude equal to the altitude of the flight destination, so the pressurization system provides a safe comfortable cabin altitude for the crew and passenger at all airplane altitude.

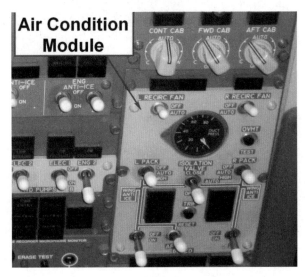

Fig.1-2　Air Conditioning Panel

7 空调系统（微课）

Air Conditioning System

Operating Principles

The air supply comes from the pneumatic system. Air is bled from the pneumatic manifold and conditioned by two separate and independent cooling packs. It is regulated by the pack valves that air flows into the packs. The air conditioning system starts after the pack valve. Air from the pneumatic mani fold is too warm and must be cooled. The function of two packs is to reduce engine bleed air temperature to the desired temperature. Basic temperature and humidity regulation are carried out in this area of the system. The cooling pack uses heat exchanger to remove heat from the air. Air conditioning is accomplished by a cooling pack utilizing the air cycle machine. A high pressure water separation system removes moisture from the air. A pack temperature controller provides automatic control of the cooling pack output temperature.

The conditioned air from the right pack flows into the main distribution manifold and the conditioned air from the left pack flows into the main distribution manifold and the flight compartment. In addition, a trim air line from the left mix chamber adds hot air to the flight compartment duct. The main distribution manifold supplies the passenger compartment overhead distribution system through two sidewall risers. A 3-phase fan draws air from the collector shrouding through filters and discharges into the main distribution manifold for redistribution.

An air cycle machine is an integral unit of each cooling pack. The air cycle machine consists of a turbine wheel and a compressor wheel mounted on a common shaft. It drives the turbine that air flows through the air cycle machine, and the turbine drives the compressor section. As the air leaves the turbine it expands greatly. This expansion can cause the temperature of the air to drop to sub-zero levels. Air cycle system installed in modern aircraft utilize air turbine refrigerating units to supply cooled air. Other model aircraft utilize a compressed gas cooling system. A freon type is the refrigerating unit. Systems utilizing this refrigeration principle are called vapor cycle system.

The air is mixed together to create a uniform temperature in the cabin. In summary, an air conditioning system is designed to perform any or all of the functions of supplying ventilation air, heated air and cooled air.

Air Conditioning Sub-systems

These are the air conditioning sub-systems:

➢ Distribution.

➢ Pressurization.

➢ Equipment cooling.

➢ Heating.

➢ Cooling.

➢ Temperature control.

Main Components

These flight compartment panels let you control the air conditioning system:

➢ Air conditioning/bleed air controls panel.

➢ Cabin temperature panel.

➢ Equipment cooling panel.

➢ Cabin pressure control panel.

Fig.1-3   3/5/6/700 Air-conditioning Panel

Fig.1-4   400/800 Air-conditioning Panel

The components in the EE compartment control the functions of the air conditioning system:

➢ Cabin temperature controller.

➢ Air conditioning accessory unit (ACAU).

➢ Cabin pressure controllers (CPCs).

Cabin Temperature Controller

The cabin temperature controller controls these functions of the air conditioning system: pack cooling temperature, flight compartment temperature and passenger cabin temperature.

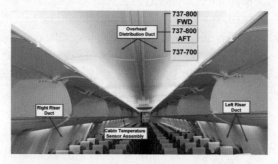

Fig.1-5   Passage Cabin (AFT View)

Air Conditioning Accessory Unit

The air conditioning accessory unit is the interface for the airplane operational logic and the air system.

Fig.1-6   ACAU

Cabin Pressure Controller

The cabin pressure controller controls the cabin pressure function of the air conditioning system.

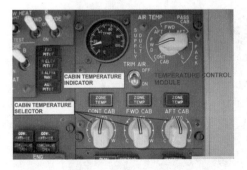

Fig.1-7   Cabin Temperature Control Module

Fig.1-8   Cabin Press Panel

Functions

The main function of an air conditioning system is to maintain a comfortable air temperature within the aircraft fuselage. The system will increase or decrease the temperature of the air as needed to obtain the desired value. Most systems are capable of producing an air temperature of 70 ℉ to 80 ℉ , with normally anticipated outside air temperatures. This temperature-conditioned air is then distributed so that there is a minimum of stratification (hot and cold layers). The system, in addition, must provide the control of humidity, it must prevent the fogging of windows, and it must maintain the temperature of wall panels and floors at a comfortable level. In a typical system, the air temperature is measured and compared to the desired setting of the temperature controls. Then, if the temperature is not correct, heaters or coolers are set into operation to change the air temperature, and the air is mixed together to create a uniform temperature in the cabin.

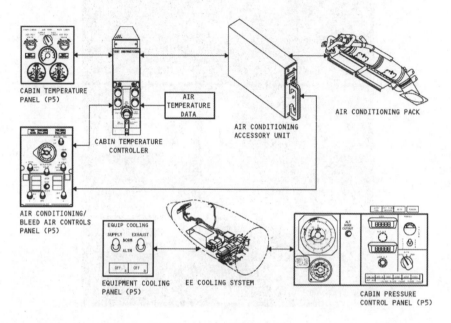

Fig.1-9　Cooling System

Warnings & Cautions

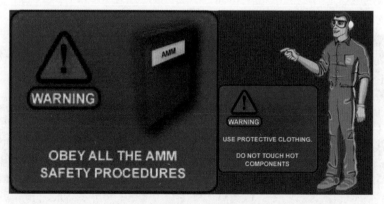

Fig.1-10　Warnings & Cautions-1

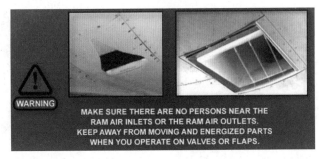

Fig.1-11   Warnings & Cautions-2

Fig.1-12   Warnings & Cautions-3

The Pressurization System

The air conditioning packs supply conditioned air into pressurization system. The pressurization system controls cabin altitude by regulating air flow from the fuselage through the outflow valve.

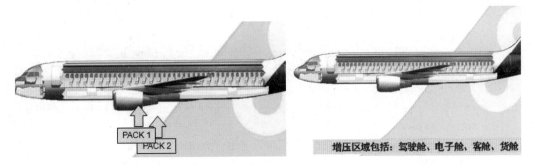

Fig.1-13   Pressurization System

The airplane starts its flight at an airport near sea level, ascends, cruises, and descends to a landing airport near sea level. Cabin altitude starts at the take off altitude. As the airplane climb, the ambient pressure decreases; the higher the altitude, the lower the pressure. The altitude inside the airplane (cabin altitude) also climbs but at a lower rate than the airplane. The pressurization system provides the pressure differential between the cabin and ambience.

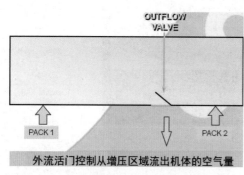

外流活门控制从增压区域流出机体的空气量

Fig.1-14　Outflow Valve-2

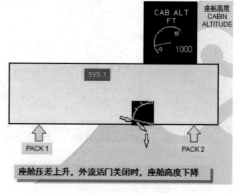

座舱压差上升，外流活门关闭时，座舱高度下降

Fig.1-15　Outflow Valve-3

The cabin pressure is referred to as "PSID" and indicates the differential pressure, in pounds per square inch, between inside and outside of the airplane. There are many systems for measuring pressure and "PSI" is only one of airplane systems measuring pressure in inches of mercury and millibars; these indications are identical to 14.7 PSI as a measurement of the pressure exerted by the atmosphere at sea level.

The pressure control system provides cabin differential pressure by controlling the outflow of air from the fuselage. It consists of aft outflow valve, forward outflow valve, pressure controller, control panel, pressure sensing inputs and monitoring indicators.

Cabin Pressure Control System

In addition, dumping of the cabin pressure is a function of the pressure control system. A cabin pressure regulator, an outflow valve and a safety valve are used to accomplish these functions. The aft outflow valve allows cabin air to exhaust overboard to maintain cabin pressure. The forward outflow valve receives a control signal from the aft outflow valve. When the aft outflow valve is closed, the forward outflow valve receives a signal to drive full closed.

Five basic requirements for the successful functioning of a cabin pressurization and air conditioning system are:

➢ A source of compressed air for pressurization and ventilation. Cabin pressurization sources can be either engine-driven compressors, independent cabin superchargers, or air bled directly from the engine.

➢ A means of controlling cabin pressure by regulating the outflow of air from the cabin. This is accomplished by a cabin pressure regulator and an outflow valve.

➢ A method of limiting the maximum pressure differential to which the cabin pressurized area will be subjected. Pressure relief valves, negative(vacuum) relief valves and dump valves are used to accomplish this.

➢ A means of regulating(in most cases cooling)the temperature of the air being distributed to the pressurized section of the airplane. This is accomplished by the refrigeration system, heat exchangers, control valves, electrical heating elements and a cabin temperature control system.

➢ The sections of the aircraft which are to be pressurized must be sealed to reduce inadvertent leakage of air to a minimum. This area must also be capable of safely withstanding the maximum pressure differential between cabin and atmosphere to which it will be subjected.

Designing the cabin to withstand the pressure differential and holding leakage of air within the limits of the pressurization system is primarily an airframe engineering and manufacturing problem.

In addition to the components just discussed, various valves, controls and allied units are necessary to complete a cabin pressurizing and air conditioning system. When auxiliary systems such as windshield rain-clearing devices, pressurized fuel tanks and pressurized hydraulic tanks are required, additional shutoff valves and control units are necessary.

Abbreviations & Acronyms

1. Air Conditioning (A/C)	空调
2. Celsius (C)	摄氏度
3. Fahrenheit (F)	华氏度
4. Full Authority Digital Engine Control (FADEC)	全权数字式发动机控制
5. Air Conditioning Accessory Unit (ACAU)	空调附件组件
6. Cabin Pressure Controller (CPC)	客舱压力控制器
7. Cabin Temperature Controller (CTC)	机舱温度控制器

Words & Expressions

1. pressurization [ˌpreʃəraɪˈzeɪʃn] *n.* 　　　增压
2. humidity [hjʊˈmɪdɪtɪ] *n.* 　　　湿气；湿度
3. buildup [ˈbɪldʌp] *n.* 　　　增大；集结
4. refrigerate [rɪˈfrɪdʒəreɪt] *vt.* 　　　使冷；使凉
5. refrigeration [rɪˌfrɪdʒəˈreɪʃən] *n.* 　　　制冷
6. pack [pæk] *n.* 　　　组件
7. landing gear bay 　　　起落架舱
8. bulkhead [ˈbʌlkhed] *n.* 　　　隔离壁
9. outflow [ˈaʊtfləʊ] *n.* 　　　流出；外流
10. pneumatic manifold 　　　气动歧管
11. trim air line 　　　配平空气管路
12. mix chamber 　　　混合室
13. sidewall riser 　　　侧壁立管
14. shroud [ʃraʊd] *n.* 　　　防护罩
15. vapor cycle system 　　　蒸汽循环系统
16. valve [vælv] *n.* 　　　活门；阀门
17. regulation [regjʊˈleɪʃ(ə)n] *n.* 　　　调节
18. accomplish [əˈkʌmplɪʃ] *vt.* 　　　完成
19. utilize [ˈjuːtɪˌlaɪz] *vt.* 　　　利用
20. freon [ˈfriːɒn] *n.* 　　　氟利昂
21. uniform [ˈjuːnɪfɔːm] *adj.* 　　　恒定的；相同的
22. panel [ˈpæn(ə)l] *n.* 　　　仪表盘；面板
23. accessory [əkˈses(ə)rɪ] *n.* 　　　配件；附件
24. anticipate [ænˈtɪsɪpeɪt] *vt.* 　　　期望
25. stratification [ˌstrætɪfɪˈkeɪʃən] *n.* 　　　分层
26. layer [ˈleɪə] *n.* 　　　层
27. differential [ˌdɪfəˈrenʃ(ə)l] *adj.* 　　　差别的；区别的

8 单词朗读

9 航空术语学习
策略（微课）

28. regulator ['reɡjʊleɪtə] *n.*　　　　　调节器；稳压器
29. relief [rɪ'liːf] *n.*　　　　　　　　释压；解除
30. negative ['neɡətɪv] *adj.*　　　　　负的
31. vacuum ['vækjʊəm] *n.*　　　　　　真空
32. leakage ['liːkɪdʒ] *n.*　　　　　　渗漏；漏出
33. ascend [ə'send] *n.*　　　　　　　上升；升高
34. descend [dɪ'send] *n.*　　　　　　下降
35. mercury ['mɜːkjəri] *n.*　　　　　水银，汞
36. millibar ['mɪlɪbɑː(r)] *n.*　　　　毫巴（大气压强单位）
37. atmosphere ['ætməsfɪə(r)] *n.*　　大气
38. ventilation [ˌventɪ'leɪʃən] *n.*　　通风
39. supercharger ['sjuːpəˌtʃɑːdʒə] *n.*　机械增压器；增压器
40. inadvertent [ˌɪnəd'vɜːtənt] *adj.*　无意的；因疏忽造成的

Exercises

2.1 Choose the best answer from A, B, C, D options.

10 答案

1. pressurization
　A. 按压　　　　　B. 压力　　　　　C. 增压　　　　　D. 释压
2. radome
　A. 雷达　　　　　B. 圆顶　　　　　C. 随机　　　　　D. 雷达罩
3. tail cone
　A. 尾椎　　　　　B. 尾部　　　　　C. 裁缝　　　　　D. 锥体
4. bulkhead
　A. 大头　　　　　B. 隔离壁　　　　C. 头顶　　　　　D. 过热
5. altitude
　A. 纬度　　　　　B. 姿态　　　　　C. 态度　　　　　D. 高度
6. 气动系统
　A. hydraulic system　　　　　　B. pressurization system
　C. pneumatic system　　　　　　D. vacuum system
7. 管道
　A. duck　　　　　B. duct　　　　　C. duchess　　　　D. duke
8. 侧壁
　A. sidewall　　　B. wallet　　　　C. walnut　　　　D. swallow
9. 过滤器
　A. file　　　　　B. fill　　　　　C. fillet　　　　D. filter
10. 蒸汽循环系统
　A. vapor cycle system　　　　　　B. ventilation control system
　C. vacuum clamping system　　　　D. air cycle machine

2.2 Find the best option to paraphrase the sentence.

1. An air conditioning system is designed to perform any or all of the functions of supplying ventilation air, heated air and cooled air.
　A. Any or all of the functions of supplying ventilation air, heated air and cooled air is performed by hydraulic system.

    B. An air conditioning system is used to supply ventilation air, heated air and cooled air.

    C. An oil system is designed to perform any or all of the functions of supplying ventilation air, heated air and cooled air.

    D. A warning system is designed to perform any or all of the functions of supplying ventilation air, heated air and cooled air.

2. The system will increase or decrease the temperature of the air as needed to obtain the desired value.

    A. The system will increase or decrease the temperature of the air as need to obtain the desired value.

    B. The system will increase or decrease the temperature of the air as required to gain the desired value.

    C. The system will increase or decrease the pressure of the air as required to develop the desired value.

    D. The system will increase or decrease the density of the air as require to get the desired value.

3. This temperature-conditioned air is then distributed so that there is a minimum of stratification (hot and cold layers).

    A. This temperature-conditioned air is then distributed so that there is a maximum of stratification (hot and cold layers).

    B. This temperature-conditioned air is then concentrated so that there is a minimum of stratification (hot and cold layers).

    C. This temperature-conditioned air is then distributive so that there is a minimum of stratification (hot and cold layers).

    D. This temperature-conditioned air is then distributed to minimize stratification (hot and cold layers).

4. This area must also be capable of safely withstanding the maximum pressure differential between cabin and atmosphere to which it will be subjected.

    A. This area must also be able to safely withstand the maximum pressure differential between cabin and atmosphere to which it will be subjected.

    B. This area must also be able to safely withstand the minimum pressure differential between cabin and atmosphere to which it will be subjected.

    C. This area must also be capable of safely withstanding the lowest pressure differential between cabin and atmosphere to which it will be objected.

    D. This area must also be capability of safely withstanding the minimum pressure between cabin and atmosphere to which it will be subjected.

5. In addition to cabin air conditioning, some aircraft equipments and equipment compartments require air conditioning to prevent heat buildup and consequent damage to the equipment.

    A. Except for cabin air conditioning, some aircraft equipments and equipment compartments don't require air conditioning to prevent heat buildup and consequent damage to the equipment.

    B. Except cabin air conditioning, some aircraft equipments and equipment components require air conditioning to get heat buildup and consequent damage to the equipment.

    C. Besides cabin air conditioning, some aircraft equipments and equipment compartments require air conditioning to prevent heat buildup and consequent damage to the equipment.

D. In addition to cabin air conditioning, some aircraft equipments and equipment compartments need air conditioning to protect heat buildup and consequent damage to the equipment.

2.3　Change the following components specified in lower case and translate them into Chinese.

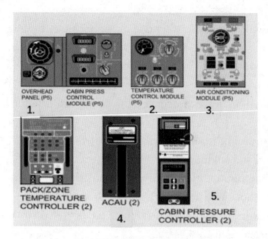

1.	
2.	
3.	
4.	
5.	

Section 3　Aviation Translation: translate the following sentences into Chinese

1. NOTE: HAVE ONE PERSON TURN THE AIR CONDITIONING PACKS ON. HAVE ONE OR MORE PERSONS CHECK THE OPERATION OF THE EMERGENCY EXIT HATCH HEATERS.

2. WARNING: MAKE SURE YOU OBEY THE PROPER PROCEDURES FOR COMPRESSION AND DECOMPRESSION WHEN YOU USE PERSONS IN A PRESSURIZED AREA. PRESSURE CHANGES THAT CAUSE PAIN MUST NOT BE DONE. IF YOU DO NOT OBEY THE PRECAUTIONS, INJURY TO PERSONS CAN OCCUR.

11 三敬精神——感动
中国 2018 年度人物
刘传健（文本）

Section 4　Aviation Writing

Liu Shuai is an aircraft mechanic. He is inspecting the plane and finds some faults.

Please help him write down fault descriptions. Some related words, phrases & terms and key sentences are offered as follows.

Key words, phrases & terms:

1. 空调舱 — air conditioning compartment
2. 空气循环机 — air cycle machine (ACM)
3. 再循环风扇 — recirculation fan
4. 出气活门 / 排气活门 — air outlet valve/ exhaust valve
5. 外溢活门 — outflow valve
6. 单向活门 — check valve
7. 关断活门 — shutoff valve
8. 释压活门 — relief valve
9. 温度控制活门 — TEMP CONT valve
10. 冲压空气作动器 — ram air actuator
11. 自动的 / 人工的 — automatic(AUTO) /manual
12. 正常的 / 备用的 — normal (NORM)/ alternate (ALTN)
13. 进气管 — inlet duct
14. 供气扇 — supply fan
15. 头顶分配管 — overhead distribution duct
16. 压力选择面板 — press selector panel
17. 客舱压力控制组件 — cabin press control module
18. 主 / 次热交换器 — primary/ secondary heat exchanger
19. 温度指示器 — temperature indicator
20. 区域温度灯 — ZONE TEMP light
21. 自动温度控制 — auto temperature control
22. 设备冷却排气关断 / 供气灯 — EQUIP cooling exhaust off/supply light

Key sentences:

1. 驾驶舱供气管道引气温度指示。
   CONT CABIN supply duct air temperature indication.
2. 前舱 / 后舱 / 驾驶舱温度自动控制。
   FWD/AFT/CONT CAB temperature auto control.
3. 无气流从脚部 / 风挡出气口流出。
   No airflow from foot /windshield outlet.
4. 人工方式下后排气活门不工作。
   Aft outflow valve does not operate in manual mode.
5. 自动失效和备用灯亮。当增压方式选择器电门在备用位时，自动失效灯灭。
   AUTO FAIL light and ALTN light come on. AUTO FAIL light goes off when pressurization mode selector switch is at the ALTN position.
6. 不能保持恰当的机舱压力。
   Can not keep the correct cabin pressure.
7. 右 PACK 灯再现时亮，警告信号抑制后熄灭。
   Right PACK light comes on during recalling and goes off after warning signal inhibited.
8. 过站检查再现故障时左组件 PACK 灯点亮，按压主警告可以熄灭。
   The L/H PACK light comes on when recalling the fault in TR, but it could extinguish by pressing master caution light.

9. 拆除并安装左发预冷器控制活门。

Remove and install the precooler control valve of L/H ENG.

10. 拆除并安装左空调组件活门。

Remove and install L/H pack valve.

Task: Liu Shuai has three writing tasks. The first one is to have a pack/zone temperature controllers Built-In-Test-Equipment test, the second is to finish removal and installation of primary heat exchanger, and the last one is to replace recirculation fan filter. Please finish the writing tasks for him.

12 警示案例——川航
3U8633- 生与死

Warning Case

Life &Death about Sichuan Airline 3U8633

Airbus A319 about Sichuan Airlines was on a flight from Chongqing to Lhasa. While the plane was cruising through Chengdu air traffic control area, the front windshield of the right seat in the cockpit of the aircraft suddenly broke and fell off, causing the cabin to be depressurized and the oxygen mask of passengers to fall off. Eventually, the aircraft landed safely at Shuangliu Airport, the co-pilot and a flight attendant were injured, and all passengers on board were safe.

Warning Tips: This incident shows the pilot's high sense of "respecting life, respecting regulations and respecting responsibility" in emergency.

Lesson 2 Flight Control System

Learning Objectives

1. Knowledge objectives:

   A. To grasp the words, related terms and abbreviations about aircraft flight control system.

   B. To grasp the key sentences about aircraft flight control system.

   C. To know the major components of flight control system.

2. Competence objectives:

   A. To be able to read and understand difficult English aviation sentences, capitalized English materials and obtain key aviation information quickly.

   B. To be able to communicate with English speakers about the topic freely.

   C. To be able to fill in job cards in English.

3. Quality objectives:

   A. To be able to self-study with the help of aviation dictionaries, Internet or other resources.

   B. To develop the craftsman spirit of carefulness and responsibility.

Section 1 Aviation Listening & Speaking

1.1 Aviation Listening: listen to the record and fill in the blanks.

1 听力录音   2 答案及原文

We will discuss the primary 1._____controls which control and balance the aircraft in 2._____longitude. No control is achieved by the use of 3._____, which are flat type controls normally fitted to the 4._____edge of the tail plane or stabilizer. These 5._____control the aircraft in pitch about the 6. _____axis. It must be adequate to balance the aircraft throughout its speed range at all permitted center of gravity positions and configurations, and to give an adequate rate of pitch for maneuvers, the pilot controls the elevators by the use of a yoke or 7._____ column. This control operates the flying control surface and is moved in four and half directed by the pilot to control the aircraft. If the pilot pulls the yoke or control column towards himself, then the elevator moves up the movement degree of the 8._____control directly affecting the angular displacement of the control surface. It produces an aerodynamic down force on the tail, which will pitch the aircraft to nose up. If the pilot pushes the yoke or stick away from himself, then the elevator will move down. Force will now be produced on the tail, which will pitch the nose of the aircraft down, varying the incidents, and hence the angle of attack of an aerofoil will also vary its lift. This system is often used instead of elevators for pitch control; the whole of the tail, plane, or stabilizer is rotated to vary its angle of attack. If the yoke is pushed forward, the leading edge of the stabilizer moves up and the trailing edge down. This increases the 9._____generated by the 10._____. So the nose of the aircraft pitches down. If the yoke is pulled back, then the stabilizer rotates in the opposite direction, and the nose pitches up.

That is the end of the lesson. You should now understand that controlling pitch is achieved by using elevators or a moving tail plane.

1.2 Aviation Speaking: practice the following dialogue and design your own.

Situation:

This is a dialogue between the pilot and the air traffic controller. They are talking about the problems about the flight control. Practice it and have your own.

Note: PIL=Pilot, GND=Ground Mechanic, CTL= Air Traffic Controller

3 对话录音

Scene 1

PIL: Guangzhou CSN 304 flaps and slats are jammed. We can't extend beyond 15, requesting divert to Zhuhai airport which has a longer runway.

CTL: CSN 304 is cleared to Zhuhai, I'll inform Zhuhai, you have a configuration problem.

4 对话译文

Scene 2

CTL: Lion 610, you are cleared to runway two-five left. Contact tower when you are in position.

PIL: Roger. Lion 610 is cleared to runway two-five left. All clear, no traffic. Let's run before takeoff.

CTL: Lion 610，fly heading 248.

PIL: Airspeed disagrees. What's going on?

CTL: Lion 610, climb to flight level 27,000 feet.

PIL: Altitude disagrees.

CTL: Lion 610, what's the nature of your problem, please?

PIL: We're experiencing a flight control problem. I have no reliable altitude information. All instruments disagree.

CTL: Roger, Lion 610. No restrictions.

PIL: It's diving! It's diving! Pull up!

Words & Expressions

1. jam [dʒæm] *v.* 卡阻，堵塞，故障
2. divert [daɪ'vɜːt] *v.* 绕道，转道
3. configuration [kən,fɪgə'reɪʃn] *n.* 构型
4. traffic ['træfɪk] *n.* 交通警告
5. airspeed ['eəspiːd] *n.* 空速
6. altitude ['æltɪtjuːd] *n.* 高度

5 课文朗读   6 课文译文

7 飞行操纵系统（微课）

Section 2 Aviation Reading

Pre-reading questions:

1. Why is a flight control system needed on an aircraft?
2. When is flight control system used?

Flight Control System

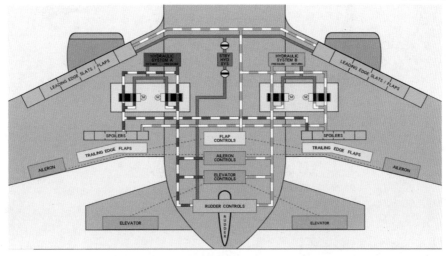

Fig.2-1 Schematic for B737-700 Flight Control System

A conventional fixed-wing aircraft flight control system consists of flight control surfaces, the respective cockpit controls, connecting linkages and the necessary operating mechanisms to control an aircraft's direction in flight. Aircraft engine controls are also considered as flight controls as they change speed.

The fundamentals of aircraft controls are explained in flight dynamics. This article centers on the operating mechanisms of the flight controls. The basic system in using on aircraft first appeared in a readily recognizable form as early as April 1908, on Louis Blériot's Blériot VIII pioneer-era monoplane design.

All flight control systems can be divided into control inputs, control transmissions and control outputs or surfaces. Control transmissions connect the control inputs with the control outputs. Control transmissions are achieved mechanically, hydraulically and electrically. The main components in a mechanical transmission are rods and a cable system. Hydraulical transmissions use hydraulic components to convert input signals into hydraulic pressure. The hydraulic pressure moves the flight control surface. Hydraulical transmissions use mechanical components as well as hydraulical components.

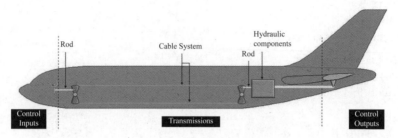

Fig.2-2　Flight Controls System

Cables must be cleaned, inspected and lubricated on a regular basis. Lint-free cloth dampened with solvent can be used to remove old lubrication grease and all dirt from the surface of the control cables.

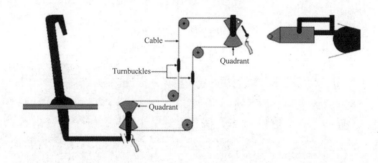

Fig.2-3　Control Cable Rigging

Fig.2-4　Cable Cleaning for Flight Control System

Cockpit Controls

Primary Controls

Generally, the primary cockpit flight controls are arranged as follows:

> A control yoke (also known as a control column), center stick or side-stick , governs the aircraft's roll and pitch. The roll is controlled by moving the ailerons when the yoke is turned or deflected left or right, and the pitch is controlled by moving the elevators when the yoke is moved backwards or forwards.

> Rudder pedals control yaw, which move the rudder. Left foot forward will move the rudder left for instance.

> Throttle controls control engine speed or thrust for powered aircraft.

Fig.2-5   A Rudder Pedal   Fig.2-6   A Control Yoke

Secondary Controls

In addition to the primary flight controls for roll, pitch and yaw, there are often secondary controls available to give the pilot finer control over flight or to ease the workload. The most commonly available control is a wheel or other device to control elevator trim, so that the pilot does not have to maintain constant backward or forward pressure to hold a specific pitch attitude (other types of trim, for rudder and ailerons, are common on larger aircraft but may also appear on smaller ones). Many aircraft have wing flaps controlled by a switch or a mechanical lever or in some cases being fully automatic controlled by computer, which alter the shape of the wing for improved control at the slower speeds used for take-off and landing. Other secondary flight control systems may be available, including slats, spoilers, air brakes and variable-sweep wings.

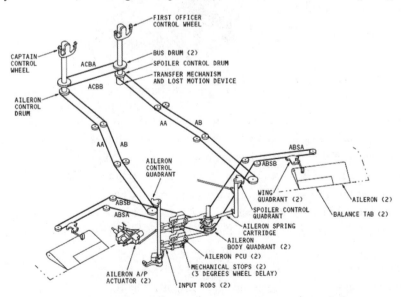

Fig.2-7   Aileron Control Trim

Fig.2-8　Slat

Fig.2-9　Spoiler

Flight Control Types

Mechanical

Mechanically or manually operated flight control systems are the most basic method of controlling the aircraft. They were used in early aircraft and are currently used in small aircraft where the aerodynamic forces are not excessive. A manual flight control system uses a collection of mechanical parts such as pushrods, tension cables, pulleys, counterweights, and sometimes chains to transmit the forces applied to the cockpit controls directly to the control surfaces. Turnbuckles are often used to adjust control cable tension. The Cessna Skyhawk is a typical example of the aircraft that uses this type of system. Gust locks are often used on parked aircraft with mechanical systems to protect the control surfaces and linkages from damage of wind. Some aircraft have gust locks fitted as part of the control system.

Fig.2-10　Cessna Skyhawk

Increase in the control surface area required by large aircraft or higher loads caused by high airspeeds in small aircraft lead to a large increase in the forces needed to move them, and consequently complicated mechanical gearing arrangements were developed to extract maximum mechanical advantage in order to reduce the forces required from the pilots. This arrangement can be found on bigger or higher performance propeller aircraft such as the Fokker 50.

Some mechanical flight control systems use servo-tabs that provide aerodynamic assistance. Servo-tabs are small surfaces hinged to the control surfaces. The flight control mechanisms move these tabs, then aerodynamic forces are in turn move, or assist the movement of the control surfaces reducing the amount of mechanical forces needed. This arrangement was used in early piston-engined transport aircraft and in early jet transports. The Boeing 737 incorporates a system, whereby in the unlikely event of total hydraulic system failure, it automatically and seamlessly reverts to being controlled via servo-tab.

The complexity and weight of mechanical flight control systems increase considerably with the size and performance of the aircraft. Hydraulically powered control surfaces help to overcome these limitations. With hydraulic flight control systems, the aircraft's size and performance are limited by economics rather than a pilot's muscular strength. At first, only-partially boosted systems were used in which the pilot could still feel some of the aerodynamic loads on the control surfaces (feedback).

A hydro-mechanical flight control system has two parts:
- ➢ The mechanical circuit, which links the cockpit controls with the hydraulic circuits. Like the mechanical flight control system, it consists of rods, cables, pulleys, and sometimes chains.
- ➢ The hydraulic circuit, which has hydraulic pumps, reservoirs, filters, pipes, valves and actuators. The actuators are powered by the hydraulic pressure generated by the pumps in the hydraulic circuit. The actuators convert hydraulic pressure into control surface movements. The electro-hydraulic servo valves control the movement of the actuators.

The pilot's movement of a control causes the mechanical circuit to open the matching servo valve in the hydraulic circuit. The hydraulic circuit powers the actuators which then move the control surfaces. As the actuator moves, the servo valve is closed by a mechanical feedback linkage - one that stops movement of the control surface at the desired position.

Fly-by-wire System

A fly-by-wire (FBW) system replaces manual flight control of an aircraft with an electronic interface. The movements of flight control are converted to electronic signals transmitted by wires (hence the fly-by-wire term), and flight control computers determine how to move the actuators at each control surface to provide the expected response. Commands from the computers are also input without the pilot's knowledge to stabilize the aircraft and perform other tasks. Electronics for aircraft flight control systems are part of the field known as avionics.

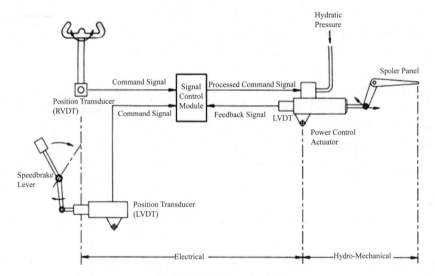

Fig.2-11　Fly-by-wire System

Modern aircraft designs like the Boeing 777 rely on sophisticated flight computers to aid and protect the aircraft in flight. These are governed by computational laws which assign flight control modes during flight. Aircraft with fly-by-wire flight control requires computer-controlled flight control modes that are capable of determining the operational mode (computational laws) of the aircraft.

A reduction of electronic flight control can be caused by the failure of a computational device, such as the flight control computer or an information providing device, such as the ADIRU. Electronic Flight Control Systems (EFCS) also provide augmentation in normal flight, such as increased protection of the aircraft from overstress or providing a more comfortable flight for passengers by recognizing and correcting for turbulence and providing yaw damping.

Fig.2-12　ADIRU

Two aircraft manufacturers produce commercial passenger aircraft with primary flight computers that can perform under different flight control modes (or laws). The most well-known are the normal, alternate, direct and mechanical laws of the Airbus A320-A380. Boeing's fly-by-wire system is used in the Boeing 777, Boeing 787 Dreamliner and Boeing 747-8.

These newer aircraft use electronic control systems to increase safety and performance while saving aircraft weight. These electronic systems are lighter than the old mechanical systems and can also protect the aircraft from overstress situations, allowing designers to reduce over-engineered components, which further reduces the aircraft's weight.

Airbus aircraft designs after the A300/A310 are almost completely controlled by fly-by-wire equipment. These newer aircraft, including the A320, A330, A340, A350 and A380, operate under Airbus flight control laws. The flight controls on the Airbus A330, for example, are all electronically controlled and hydraulically activated. Some surfaces, such as the rudder, can also be mechanically controlled. In normal flight, the computers act to prevent excessive forces in pitch and roll.

Abbreviations & Acronyms
1. Fly-by-wire System (FBW)　　　　　　　电传操纵系统
2. Electronic Flight Control Systems (EFCS)　电子飞行控制系统
3. Hydro-mechanical Unit(HMU)　　　　　液压机械组件

4. Air Data/Inertial Reference Unit (ADIRU)　　大气数据惯性组件

Words & Expressions

1. flight control system　　飞行操纵系统
2. connecting linkage　　连杆机构
3. dynamic [daɪˈnæmɪk] *n.*　　动态，动力
4. mechanism [ˈmek(ə)nɪz(ə)m] *n.*　　机械装置，机构
5. monoplane [ˈmɒnəpleɪn] *n.*　　单翼机
6. transmission [trænzˈmɪʃ(ə)n] *n.*　　传动装置，变速器
7. rod [rɒd] *n.*　　杆
8. cable [keɪb(ə)l] *n.*　　钢索
9. lubricate [luːbrɪkeɪt] *v.*　　润滑
10. lint [lɪnt] *n.*　　棉绒；毛绒
11. solvent [sɒlv(ə)nt] *n.*　　溶剂
12. grease [griːs] *n.*　　油脂
13. control yoke　　操作杆
14. deflect [dɪˈflekt] *v.*　　使转向；使偏斜
15. pedal [ˈpedl] *n.*　　踏板
16. roll [rəʊl] *n.*　　翻滚
17. pitch [pɪtʃ] *n.*　　俯仰
18. yaw [jɔː] *n.*　　偏航
19. throttle [ˈθrɒt(ə)l] *n.*　　节流阀，油门
20. ease [iːz] *n.*　　缓解
21. workload [ˈwɜːkləʊd] *n.*　　工作量
22. variable-sweep wing　　可变后掠翼
23. turnbuckle [tɜːnbʌkl] *n.*　　套筒螺母
24. pushrod [ˈpʊʃrɒd] *n.*　　推杆
25. tension cable　　张力索
26. pulley [ˈpʊli] *n.*　　滑轮
27. counterweight [ˈkaʊntəweɪt] *n.*　　配重；平衡物
28. propeller [prəˈpelə(r)] *n.*　　螺旋桨
29. servo-tab [ˈsɜːvəʊ tæb]　　伺服调整片
30. incorporate [ɪnˈkɔːpəreɪt] *v.*　　合并；包含
31. seamlessly [ˈsiːmləs] *adj.*　　无缝地
32. revert to　　恢复
33. aerodynamic load　　气动力负载
34. circuit [ˈsɜːkɪt] *n.*　　线路，电路
35. interface [ˈɪntəfeɪs] *n.*　　界面，接口
36. hydraulic pump　　液压泵
37. reservoir [ˈrezəvwɑː(r)] *n.*　　储液罐
38. actuator [ˈæktjʊeɪtə] *n.*　　作动筒，致动器
39. electronic control system　　电子操控系统
40. fly-by-wire flight control　　电传飞行操控
41. augmentation [ˌɔːgmɛnˈteɪʃ(ə)n] *n.*　　增强
42. turbulence [ˈtɜːbjələns] *n.*　　湍流

8 单词朗读

9 术语专讲：
蹬舵（微课）

25

43. yaw damping   偏航阻尼
44. law [lɔ:] *n.*   准则；规矩
45. computational law   计算定律

10 答案

Exercises

2.1 Choose the best answer from A, B, C, D options.

1. throttle
   - A. 升力
   - B. 油门
   - C. 推力
   - D. 阻力
2. pulley
   - A. 滑轮
   - B. 滑冰
   - C. 滑雪
   - D. 拉力
3. counterweight
   - A. 超重
   - B. 配平
   - C. 净重
   - D. 配重
4. avionic
   - A. 航空
   - B. 航空电子
   - C. 电子
   - D. 飞行
5. air brake
   - A. 加速
   - B. 超速
   - C. 减速板
   - D. 失速
6. 钢索
   - A. duct
   - B. wire
   - C. rod
   - D. cable
7. 偏航
   - A. roll
   - B. yaw
   - C. pitch
   - D. tab
8. 襟翼
   - A. aileron
   - B. slat
   - C. flap
   - D. rudder
9. 脚踏板
   - A. pedal
   - B. panel
   - C. column
   - D. yoke
10. 润滑
    - A. lubrication
    - B. grease
    - C. liquid
    - D. fluid

2.2 Find the best option to paraphrase the sentence.

1. Aircraft engine controls are also considered as flight controls as they change speed.
   - A. Aircraft engine controls are also considered as flight controls as they don't change speed.
   - B. Aircraft engine controls are not considered as flight controls although they change speed.
   - C. Aircraft engine controls are also regarded as flight controls as they change speed.
   - D. Aircraft engine controls are also known as flight controls when they change speed.
2. Mechanically or manually operated flight control systems are the most basic method of controlling an aircraft.
   - A. Mechanically or automatically operated flight control systems are the most basic method of controlling an aircraft.
   - B. The most basic method of controlling an aircraft is mechanically or manually operated flight control systems.
   - C. Mechanical or manual operation of flight control systems is the most simple method of controlling an aircraft.
   - D. Mechanical or electronic operation of flight control systems is the most important method of controlling an aircraft.

3. This article centers on the operating mechanisms of the flight controls.
   A. This article focuses on the operating mechanisms of the flight controls.
   B. The importance of this article is put on the mechanical operation of the flight controls.
   C. This article's core is the operation of machinery of the flight controls.
   D. This article doesn't center on the operation of the flight controls.

4. Modern aircraft designs like the Boeing 777 rely on sophisticated flight computers to aid and protect the aircraft in flight.
   A. Modern aircraft designs like the Boeing 777 rely on simple flight computers to aid and protect the aircraft in flight.
   B. Modern aircraft designs like the Boeing 777 rely on sophisticated flight computers to aid and prevent the aircraft in flight.
   C. Modern aircraft designs like the Boeing 777 rely on sophisticated flight computers to control the aircraft in flight and during landing.
   D. Modern aircraft designs like the Boeing 777 depend on sophisticated flight computers to aid and protect the aircraft in flight.

5. A fly-by-wire(FBW) system replaces manual flight control of an aircraft with an electronic interface.
   A. A fly-by-wire(FBW) system exchanges manual flight control of an aircraft with an electrical interface.
   B. A fly-by-wire(FBW) system takes place manual flight control of an aircraft with an electronic interface.
   C. A fly-by-wire(FBW) system takes the place of manual flight control of an aircraft with an electronic interface.
   D. A fly-by-wire(FBW) system takes care of manual flight control of an aircraft with an electronic interface.

2.3 Change the following components specified in lower case and translate them into Chinese.

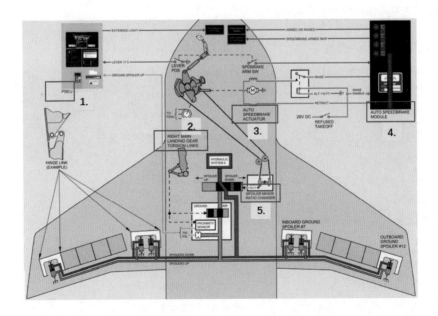

1.	
2.	
3.	
4.	
5.	

Section 3 Aviation Translation: translate the following sentences into Chinese

1. WARNING: MAKE SURE THAT PERSONS AND EQUIPMENT ARE CLEAR OF ALL CONTROL SURFACES BEFORE YOU SUPPLY HYDRAULIC POWER. AILERONS, RUDDER, ELEVATORS, FLAPS, SPOILERS, LANDING GEAR, AND THRUST REVERSERS CAN MOVE QUICKLY WHEN YOU SUPPLY HYDRAULIC POWER. THIS CAN CAUSE INJURIES TO PERSONS AND DAMAGE TO EQUIPMENT.

11 岗位自豪（文本）

2. WARNING: MAKE SURE THAT THE POSITION OF THE TE FLAPS AGREES WITH THE POSITION OF THE FLAP CONTROL LEVER. WHEN YOU SUPPLY HYDRAULIC POWER, THE FLAPS AND SLATS WILL MOVE AUTOMATICALLY TO THE POSITION OF THE FLAP CONTROL LEVER.

Section 4 Aviation Writing

Liu Shuai is an aircraft mechanic. He is inspecting the plane and finds some faults.

Please help him write down fault descriptions. Some related words, phrases & terms and key sentences are offered as follows.

Key words, phrases & terms:

1. 减速板   airbrake
2. 机翼下副翼配平   wing down aileron trim
3. 副翼 / 方向舵配平作动器   aileron/ rudder trim actuator
4. 襟翼位置指示器   flap position indicator
5. 襟翼收放测试   test for R/E（retract/ exten）flaps
6. 控制杆 / 轮   control column/ wheel
7. 巡航   cruise
8. 接近   approach
9. 机头向上 / 下   nose up/ down
10. 转换机构   transfer mechanism
11. 驱动电动 / 液压马达   drive electric/ hydraulic motor
12. 失速警告测试面板   stall warning test panel
13. 内锁 / 旁通活门   interlock/ bypass valve
14. 配平空气活门   trim air valve

15. 自动缝翼控制活门　　　　auto-slat control valve

16. 操纵台泛光灯　　　　　　control stand flood light

17. 减速板预位灯　　　　　　speed brake armed light

18. 减速板不预位灯　　　　　speed brake do not arm light

19. 前缘缝翼过渡灯　　　　　leading edge slats transit light

Key sentences of direction description:

1. 襟翼过渡灯持续亮。

   FLAPS TRANSIT light stays on.

2. 襟翼手柄不能活动自如。

   Flap lever does not move freely.

3. 副翼对驾驶盘反应慢。

   Aileron showed slow response to control wheel movement.

4. 左机翼向下和右机翼向下不工作。

   Left wing down and right wing down did not operate.

5. 操作测试飞控关断活门。

   Do an operational test of flight control shutoff valves.

6. 飞行卡位不工作。

   Flight detent does not operate.

7. 手柄不能灵活移动。

   Lever does not move freely.

8. 扰流板不能在一个方向上移动。

   Spoilers do not move in one direction.

9. 扰流板面板飘浮。

   Spoiler panel(s) float.

10. 润滑地面扰流板内锁活门钢索。

    Lubricate ground spoiler interlocks valve cable.

11. 拆装 4 号地面扰流板外侧作动器。

    Remove and install No.4 ground spoiler outboard actuator.

12. 润滑扰流板比例混合器。

    Lubricate spoiler ratio changer.

13. 操作测试扰流板关断活门。

    Do an operational test of spoiler shutoff valves.

Task: Liu Shuai finds out three problems and tasks in the check. The first one is that the flight detent does not work, the second is to remove and install No.4 ground spoiler outboard actuator, and the last one is to have an operational test of flight control shutoff valves. Please finish the writing tasks for him.

12 警示案例——狮子航空 610 雏鹰折翅

Warning Case

The Disaster of Lion Air Flight 610

Indonesian Lion Airlines Air Flight 610 flew from Jakarta to Penang on the morning of October 29, 2018, but crashed into the Java Sea shortly after taking off. There were 189 people on board.

Warning Tips: the MCAs (Maneuver Enhancement System) Augndation system caused the aircraft crash. The system is designed to help prevent the aircraft from stalling, but due to the wrong data of the sensor triggered by mistakes, the aircraft has entered an unrecoverable dive state.

Lesson 3　Fuel System

Learning Objectives

1. Knowledge objectives:

   A. To grasp the words, related terms and abbreviations about fuel system on the aircraft.

   B. To grasp the key sentences about fuel system.

   C. To know the major components of fuel system.

2. Competence objectives:

   A. To be able to read and understand frequently-used & complex sentence patterns, capitalized English materials and obtain key information quickly.

   B. To be able to communicate with English speakers freely.

   C. To be able to fill in job cards in English.

3. Quality objectives:

   A. To be able to self-study with the help of aviation dictionaries, the Internet and other resources.

   B. To develop the craftsman spirit of carefulness and responsibility.

Section 1　Aviation Listening & Speaking

1.1 Aviation Listening: listen to the record and fill in the blanks.　1 听力录音　2 答案及原文

　　Have you ever asked yourself how often an airliner 1._____? Do airliners only carry so much 2._____for one flight or more flights? You immediately think about your maximum 3._____of your hand luggage you are allowed to bring on your flight. So why would a plane carry tons of unnecessary fuel around? That will be the topic of today.

　　I'll be explaining you the fuel 4._____tankering procedure which is a common 5._____ on shorter flights where the maximum 6._____and landing weight aren't an issue. So I'll be going through the fuel 7._____procedure step by step. I'll be mentioning the 8._____fuel, trip fuel, contingency fuel, alternate fuel and final reserve. Which of these figures are more important or less during your fuel planning. Finally, I'll be explaining you why the fuel profit tankering is not the most economically friendly 9._____, but one how airlines become more profitable and use this technic to 10._____fuel.

I hope you enjoy this tutorial on fuel tankering and use this procedure in your upcoming flights.

1.2 Aviation Speaking: practice the following dialogue and design your own.

Situation:

This is a dialogue between the pilot and the ground mechanic. After startup, leaking fuel on the ground is found. Practice it and have your own.

Note: PIL=Pilot, GND=Ground Mechanic

3 对话录音

Scene 1

GND: Cockpit, Ground.

PIL:　Go ahead.

GND: The engine NO.1 leak fuel during start. Keep the engine NO.1 in idle and Standby. I need to access the engine and do some inspection. I need to report the fault to the engineer now.

4 对话译文

PIL:　Roger, keep the engine NO.1 in idle and Standby.

Scene 2

GND: Cabin, shut down the left engine now! The fuel is leaking out from the engine.

PIL:　Roger. Shutting down. What does the leak look like?

GND: It looks like a serious problem. We need to tow the plane back for the further inspections.

Are there any abnormal phenomena shown in the cabin?

PIL:　Not yet, the FWC (flight warning computer) shows nothing on the display. And what do you think the delay for this flight? How long will it take for the further inspections?

GND: I am not sure yet, it depends on how serious and where the leakage problem is. The plane needs further inspections. It's hard to make any judgement right now.

PIL:　OK, understand. Thanks. Waiting for you further information.

GND: OK. If there is any further maintenance information, we will inform the SOC as soon as possible.

PIL:　Understand, we are going to contact the SOC to see if they can arrange another flight or change the plane for us. Thanks.

Words & Expressions

1. Security Operations Center（SOC）　　安全运行中心
2. idle ['aɪdl] *adj.*　　慢车的
3. access ['ækses] *v.*　　进入

Section 2　Aviation Reading

Pre-reading questions:

1. Why does the aircraft need fuel?

5 课文朗读

6 课文译文

Fuel System

7 发动机燃油
（微课）

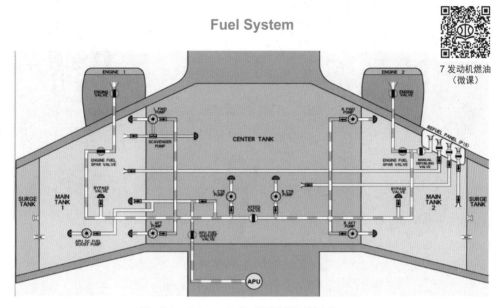

Fig.3-1　Schematic for B737-700 Fuel System

Introduction

An aircraft fuel system allows the crew to pump, manage, and deliver aviation or jet fuel to the propulsion system and Auxiliary Power Unit (APU) of the aircraft. Fuel systems differ greatly due to different performance of the aircraft in which they are installed. A single-engine piston aircraft has a simple fuel system; a tanker (such as the KC-135), in addition to managing its own fuel, can also provide fuel to other aircraft.

Fuel is piped through the fuel lines to a fuel control valve (usually known as the fuel selector). This valve serves several functions. The first function is to act as a fuel shutoff valve. This is required to provide the crew with a means to prevent fuel reaching the engine in case of an engine fire. The second is to allow the pilot to choose which tank feeds the engine. Many aircraft have the left tank and right tank selections available to the pilot. Some Cessna airplanes feed only from both tanks, and many have the option to feed from left, right, or both tanks. In some aircraft, the shutoff function is a different valve located after the fuel selector valve.

Typically, after the selector valve-situated at a low point in the fuel run-there is a gascolator-a fuel filter that can be opened on the ground and drained of fuel impurities denser than petroleum, mainly water and sediment. Other drainage points are in each tank (often more than one contaminant collection sump per tank) and at the injection pump.

Each tank is vented (or pressurized) to allow air into the tank to take the place of burned fuel; otherwise, the tank would be in negative pressure which would result in engine fuel starvation. A vent also allows for changes in atmospheric pressure and temperature.

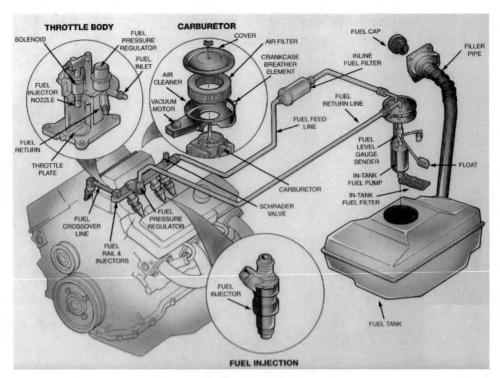

Fig.3-2　Components of Fuel System

Turbine Fuel System

All of the considerations made for the twin piston are applicable to turbine fuel system. Additional consideration applies because of the higher altitudes, different fuel, lower temperatures and longer flights.

To avoid water condensation or the fuel itself solidifying at low temperatures (-55℃), fuel tanks have thermometers and heating systems. Many are pressurized with engine bleed air to keep moist air out and ensure positive pressure feed to the pumps. In larger aircraft, fuel tanks are also in the fuselage and their load affects the center of gravity of the aircraft. This imposes limitations on the amount of fuel carried and the order in which fuel must be used. Turbine engines burn fuel faster than reciprocating engines do. Because fuel needs to be injected in to a combustor, the injection system of a turbine engine aircraft must provide fuel at higher pressure and flow compared to that for a piston engine aircraft.

The refueling system of larger aircraft includes a single positive pressure refueling point from which all tanks can be fueled. How much and to which tanks fuel is fed during refueling operations is determined by the controls in the refueling panel, usually installed nearby and accessible to ground crews.

Subsystems of Fuel System

The fuel system has the primary purposes of storing fuel for use by engines and APU, supplying fuel to engines and to APU. It is subdivided into storage, fueling, distribution and indicating systems. It has these subsystems of fuel storage, pressure fueling, engine fuel feed, APU fuel feed, defuel, fuel quantity indicating system and fuel temperature indication.

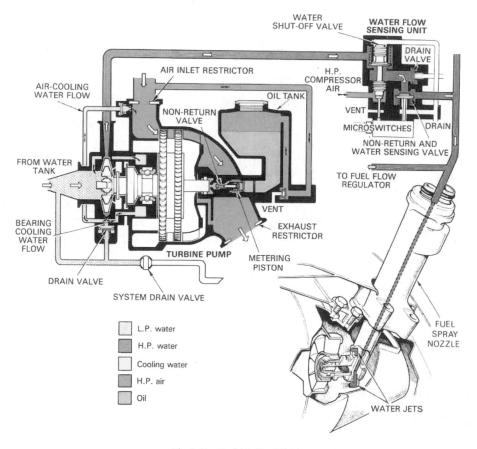

WATER
SHUT-OFF VALVE

WATER FLOW
SENSING UNIT

DRAIN
VALVE

AIR INLET RESTRICTOR

H.P.
COMPRESSOR
AIR

AIR-COOLING
WATER FLOW

OIL TANK

NON-RETURN
VALVE

VENT

MICROSWITCHES   DRAIN

NON-RETURN AND
WATER SENSING VALVE

FROM WATER
TANK

TO FUEL FLOW
REGULATOR

BEARING
COOLING
WATER
FLOW

VENT

EXHAUST
RESTRICTOR

DRAIN VALVE

TURBINE PUMP

METERING
PISTON

SYSTEM DRAIN VALVE

FUEL
SPRAY
NOZZLE

L.P. water

H.P. water

Cooling water

H.P. air

Oil

WATER JETS

Fig.3-3   Turbine Fuel System

Fuel Storage System

There are three tanks for fuel storage and two vent surge tanks for temporary fuel storage. Tank No.1 is located in the left wing, No.2 is located in the right wing and center tank is located in the fuselage under the passenger cabin (in the fuselage and the inboard section of each wing). The two vent surge tanks are located outboard of the main tanks. All fuel tanks are fuel tight. Sealing compounds and sealed fasteners are used on all joints to complete the fluid tight seal. Two of the wing ribs contain a series of baffle check valves to prevent fuel flow away from the boost pumps.

The fueling system provides a means of filling the fuel tanks. Fueling can be accomplished by the use of a single pressure fuel station through which all tanks can be filled partially or completely. Alternately, the main tanks (No.1 and No.2) can be filled through the over wing ports. The center tank can be filled by transferring fuel from the main tanks using the fuel station.

The components associated with the fueling system are the fuel station located under the right wing leading edge outboard of the engine, three float switches (one for each tank), to prevent the tanks from being overfilled. The tank fueling float switches are located near the top of each tank and sense full tank quantity. The tanks can be filled to any desired quantity; the float switches sense full tank condition and automatically stop the fueling. Power requirements are supplied by ground power, APU or the battery. Pressure fueling can also be done manually. Maximum fuel pressure is 55 PSI.

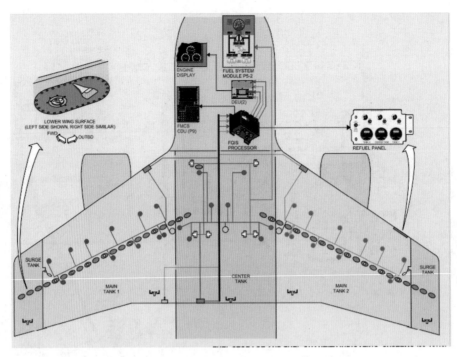

Fig.3-4　Fuel Storage & Quantity Indicating Systems

Distribution System

The distribution system allows fuel to be supplied to the engines and APU. In addition, the system can be used for defueling the tanks and for fuel transfer between the tanks. The distribution system utilizes pumps, valves and turbines for engine and APU feed. The fuel station is used for defueling and transfer operations which are only possible on the ground.

Fuel is first supplied to both engines from the center tank and then from the respective tanks to engines. The cross feed valve allows fuel from one tank to be supplied to both engines. Fuel to the APU is primarily supplied from tank No.1, but it can be supplied from any tank.

The fuel quantity indicating units measure the weight of usable fuel in the tanks. Two types of quantity indicators are used. One is capacitance indication, and the other is a manual measuring stick.

Digital fuel quantity indicators show the weight of fueling each tank. Each indicator consists of a single chip microcomputer system and a digital Liquid Crystal Display(LCD). The microcomputer measures the capacitance and resistive current in the tank units. It monitors the leakage current. When it reaches an unacceptable level, a code 0 through 10 illuminates to assist in troubleshooting the system. The measuring sticks are graduated tape which can be pulled down for reading of fuel height.

The fuel tanks store fuel for use by the engines and the APU. The pressure fueling system lets you add fuel to each tank. The fuel station is on the right wing. You also do defueling and fuel transfer at the fuel station.

Each main tank has two boost pumps (fuel pumps). The center tank also has two boost pumps. The center tank boost pumps supply fuel at a higher pressure than the pumps in the main tanks. Because of this, the fuel in the center tank is used before the fuel in the main tanks. Control of the engine and APU fuel feed system is on the P 5 panel. Fuel quantity of each tank is showed in the flight compartment and at the fuel station. BITE is available to maintenance personnel through the Control Display Unit (CDU).

Pressure Fueling System

The pressure refueling system fuels each fuel tank. The P 15 fueling panel on the right wing controls fueling operations. There is no over wing fueling capability.

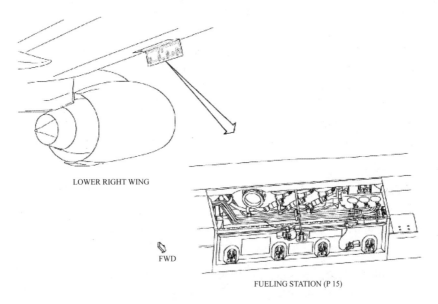

LOWER RIGHT WING

FWD

FUELING STATION (P 15)

Fig.3-5　Pressure Fueling System

Engine Fuel Feed System

The engine fuel feed system supplies fuel to the engines from main tank No.1, main tank No.2 and the center tank. The engines use fuel from the center tank before the main tanks. Operate the engine fuel feed system from the fuel control panel (P 5-2) and the engine start levers (P 8).The fuel control panel controls engine fuel feed.

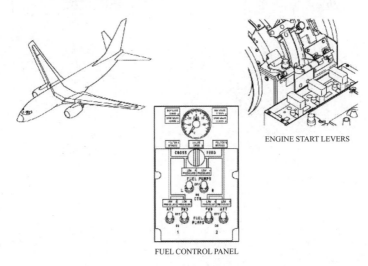

ENGINE START LEVERS

FUEL CONTROL PANEL

Fig.3-6　Engine Fuel Feed System

APU Fuel Feed System

The APU fuel feed system supplies fuel to the APU. The APU usually receives fuel from main tank 1. However, with use of the fuel boost pump switches, any fuel tank can supply fuel to the APU. The center tank boost pumps or the boost pumps in main tank No. 1 and main tank No. 2 supply fuel to the APU. If the boost pumps are off, the APU suctions fuel from main tank No. 1. ADC boost pump supplies fuel to the APU when the center and main tank boost pumps cannot supply fuel to the left fuel feed manifold. The Electronic Control Unit (ECU) controls fuel flow to the APU.

The ECU receives inputs from these items:

➢ APU master switch.
➢ Fire protection system.
➢ APU sensors.

The ECU uses these inputs to control the APU fuel shutoff valve. The fuel shutoff valve battery makes sure that the fuel system always has power to close the APU fuel shutoff valve.

Defuel System

The defuel system permits pressure defuel of each tank and suction defuel of main tank No. 1 and main tank No. 2. It also allows fuel transfer on the ground from one fuel tank to another. Use refuel station, fuel pumps, defuel valve and cross feed valve to pressure defuel the tanks. Use the defuel valve and refuel station to suction defuel main tank No. 1 and main tank No. 2.

Fuel Quantity Indicating System(FQIS)

The FQIS shows fuel weight of the main tanks and the center tank on the Common Display System (CDS) and the P 15 refuel panel. Total fuel weight shows in the Flight Management Computer System (FMCS) data on the CDU.

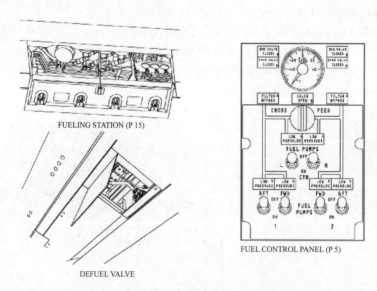

FUELING STATION (P 15)

DEFUEL VALVE

FUEL CONTROL PANEL (P 5)

Fig.3-7　Defuel System

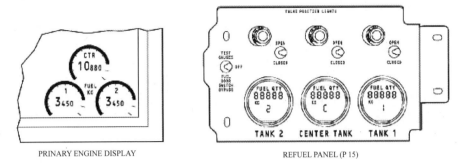

PRINARY ENGINE DISPLAY   REFUEL PANEL (P 15)

Fig.3-8   FQIS

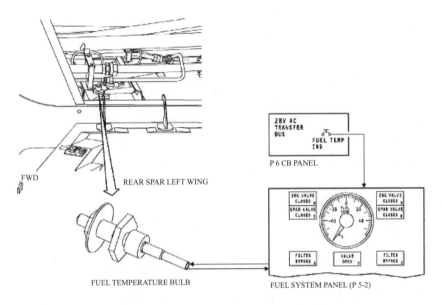

FWD   REAR SPAR LEFT WING

FUEL TEMPERATURE BULB   FUEL SYSTEM PANEL (P 5-2)

Fig.3-9   Fuel Temperature Indicating System

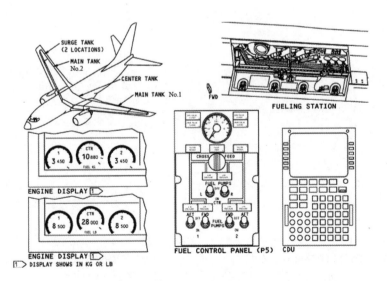

Fig.3-10   Fuel General Description

Abbreviations & Acronyms

1. Common Display System (CDS)　　　　　公共显示系统
2. Fuel Quantity Processor Unit (FQPU)　　燃油量处理组件
3. Fuel Quantity Indicating System (FQIS)　燃油量指示系统
4. Kilograms (Kgs)　　　　　　　　　　　　千克
5. Flight Management Computer System (FMCS)　飞行管理计算机系统
6. Pounds (Lbs)　　　　　　　　　　　　　磅
7. PSI (Pound per Square Inch)　　　　　　磅 / 平方英寸
8. Liquid Crystal Display (LCD)　　　　　　液晶显示器

8 单词朗读

9 术语专讲：空中加油（微课）

Words & Expressions

1. deliver [dɪˈlɪvə(r)] n.　　　　　　　传送，传输
2. tanker [ˈtæŋkə(r)] n.　　　　　　　罐车；油槽车
3. fuel line　　　　　　　　　　　　　燃油管路
4. feed [fiːd] v.　　　　　　　　　　　供给，供油
5. impurity [ɪmˈpjʊərəti] adj.　　　　杂质
6. dense [dens] adj.　　　　　　　　　稠密的
7. petroleum [pəˈtrəʊliəm] n.　　　　石油；原油
8. sediment [ˈsedɪmənt] n.　　　　　沉淀物
9. contaminant [kənˈtæmɪnənt] n.　　污染物
10. injection pump　　　　　　　　　　喷油泵
11. negative pressure　　　　　　　　　负压
12. twin piston　　　　　　　　　　　　双活塞
13. reciprocating engine　　　　　　　往复式发动机
14. vent [vɛnt] n.　　　　　　　　　　通气装置
15. defuel [diːˈfjuə] v.　　　　　　　　排油，抽油
16. sealing compound　　　　　　　　　密封剂
17. joint [dʒɔɪnt] n.　　　　　　　　　结合处；接头
18. baffle [ˈbæfl] n.　　　　　　　　　缓冲隔板
19. check valve　　　　　　　　　　　　单向活门
20. boost pump　　　　　　　　　　　　增压泵
21. fuel station　　　　　　　　　　　　加油口，加油站
22. float switch　　　　　　　　　　　　浮子电门；浮子开关
23. cross feed valve　　　　　　　　　　交输（供油）活门
24. capacitance [kəˈpæsətns] n.　　　　电容量
25. chip [tʃɪp] n.　　　　　　　　　　　芯片
26. resistive current　　　　　　　　　　阻性电流，电阻电流
27. leakage current　　　　　　　　　　泄露电流
28. graduated [ˈɡrædʒʊetɪd] adj.　　　有刻度的
29. vent surge tank　　　　　　　　　　通风（防震动）油箱
30. tight [taɪt] adj.　　　　　　　　　　密封的
31. surge [sədʒ] v.　　　　　　　　　　波动，喘振
32. temporary [ˈtempəreri] adj.　　　　临时的
33. sealing [ˈsiːlɪŋ] n.　　　　　　　　　密封
34. compound [kɒmˈpaʊnd] n.　　　　　化合物

35. fastener ['fæsnə] *n.*   紧固件
36. alternately ['ɔltənɪtli] *adv.*   交替地
37. cross [krɔs] *v.*   交输
38. quantity indicating   数量指示器
39. stick [stɪk] *n.*   操纵杆，手柄
40. microcomputer ['maɪkrəʊ-kəm'pjʊtə] *n.* 微型计算机
41. leakage ['likɪdʒ] *v/n.*   泄漏；渗漏
42. code [kɒd] *n.*   代码

Exercises

2.1 Choose the best answer from A, B, C, D options.

1. fuel
   A. 燃油   B. 滑油   C. 液压油   D. 油脂

10 答案

2. tanker
   A. 坦克   B. 油槽车   C. 加油机   D. 燃油

3. thermometer
   A. 高度表   B. 温度计   C. 速度表   D. 陀螺仪

4. chip
   A. 芯片   B. 便宜   C. 炸薯条   D. 吱吱叫

5. capacitance
   A. 能力   B. 容积   C. 电阻   D. 电容

6. 关断活门
   A. control valve   B. check valve   C. cross feed valve   D. shutoff valve

7. 喷射泵
   A. fuel pump   B. injection pump   C. boost pump   D. water pump

8. 加油
   A. refuel   B. storage   C. distribution   D. indication

9. 翼肋
   A. wing spar   B. winglet   C. wingtip   D. wing rib

10. 公共显示系统
   A. Content Delivery Servicer(CDS)   B. Control and Display System(CDS)
   C. Common Display System (CDS)   D. Cold Drawn Steel(CDS)

2.2 Find the best option to paraphrase the sentence.

1. Fuel systems differ greatly due to different performance of the aircraft in which they are installed.
   A. Fuel systems are not very different because they are installed on aircraft with different performance.
   B. Fuel systems are different greatly because they have different performance.
   C. Fuel systems vary widely because they are installed on aircraft with different performance.
   D. Fuel systems are the same due to different performance of the aircraft in which they are installed.

2. Fuel is piped through fuel lines to a fuel control valve.
   A. Fuel is delivered to a fuel control valve through the fuel lines.

B. Fuel is not being fed through the fuel lines to a fuel control valve.

C. A fuel control valve delivers fuel to the fuel lines.

D. Fuel is delivered to a fuel control valve of the fuel lines.

3. Turbine engines burn fuel faster than reciprocating engines do.

A. Reciprocating engines use fuel faster than turbine engines do.

B. Turbine engines consume fuel faster than reciprocating engines do.

C. Turbine engines use fuel at the same speed as reciprocating engines do.

D. Turbine engines do not burn fuel faster than reciprocating engines do.

4. How much and to which tanks fuel is fed is determined by the controls in the refueling panel.

A. The controls in the refueling panel don't determine the amount of refueling and which tank to refuel.

B. The controls in the refueling panel can only determine the amount of refueling.

C. The controls in the refueling panel cannot decide which tank to refuel.

D. The controls in the refueling panel determine the amount of refueling and which tank to refuel.

5. The fuel quantity indicating units measure the weight of usable fuel in the tanks.

A. The fuel quantity indicating units measure the remaining fuel in the tanks.

B. The quantity of usable fuel in the tanks is measured by the fuel quantity indicating units.

C. The fuel quantity indicating units measure the unusable fuel in the tanks.

D. The fuel quantity indicating units measure fuel which has been used.

2.3 Change the following components specified in lower case and translate them into Chinese.

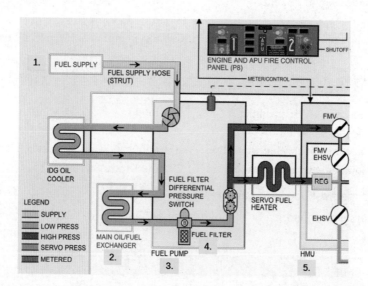

1.	
2.	
3.	
4.	
5.	

Section 3 Aviation Translation: translate the following sentences into Chinese

1. NOTE: THE EQUIPMENT IS NOT FIELD ADJUSTABLE. THE CALIBRATION MUST BE DONE IN A MAINTENANCE SHOP.

2. CAUTION: DO NOT USE WIDE CUT FUEL WHEN IT IS NOT PERMITTED. A FLAMEOUT CAN OCCUR AND ENGINE POWER CAN DECREASE SUDDENLY.

11 精致维修——机务
团队的"工匠精神"
（文本）

Section 4 Aviation Writing

Liu Shuai is an aircraft mechanic. He is inspecting the plane and finds some faults.

Please help him write down fault descriptions. Some related words, phrases & terms and key sentences are offered as follows.

Key words, phrases & terms:

1. 燃油传输   fuel transfer
2. 中央油箱   center tank
3. 燃油泵油箱   fuel pump tank
4. 加油喷嘴   refuel nozzle
5. 加油指示器   fueling indicator
6. 大翼加油面板   wing fueling panel
7. 加油浮子电门   refuel float switch
8. 燃油泵低压灯   fuel pump LOW PRESSURE light
9. 测试油量表   TEST GAGES
10. 燃油交输活门   cross feed valve
11. 主油箱   main tank
12. 通风油箱   surge tank
13. 放油   defuel
14. 沉积物   sediment
15. 火焰抑制器   flame arrester
16. 补偿器   compensator
17. 加油站   refuel station
18. 燃油通风口   fuel vent
19. APU 供油   APU feed
20. 燃油量   fuel quantity
21. 加油面板燃油量指示器   fuel quantity indicator on refuel panel

Key sentences:

1. 显示指示器失效。
   Shows IND FAIL.
2. 增压泵跳开关拔出。
   Boost pump circuit breaker opens.
3. 电门在测试油量表位时，显示不正确。
   Display is not correct with switch at TEST GAGES.

4. 飞机接通地面电源时加油面板灯不亮。

Refueling panel light does not come on with ground power on the airplane.

5. N1，N2，EGT 及燃油流量指示读数低或波动。

N1, N2, EGT and FF indications read low or fluctuate.

6. 更换燃油滤滤筒。

Replace fuel filter cartridge.

1）波动太高或太低。

Fluctuation is too high or too low.

2）闪亮。

Flashes.

3）操作测试燃油量指示系统。

Do the operational test of fuel quantity indicating system.

7. 燃油交输活门开灯。

Fuel cross feed VALVE OPEN light.

1）活门在过渡过程时，不亮。

Does not come on when the valve is in transit.

2）交输选择器移至关断位时，保持明亮。

Stay on bright when the cross feed selector is moved to the off position.

3）交输选择器移至接通位时，保持明亮。

Stay on bright when the cross feed selector is moved to the on position.

8. 加油活门灯。

Refuel valve light.

1）所选活门打开且加油总管增压时，灯不亮。

Light does not come on with valves selected open and refueling manifold pressurized.

2）按压"按压测试"电门后，灯不灭。

Light does not go off after pressing the press-to-test switch.

3）所选活门打开且加油总管增压时，加油活门位置指示器灯不亮。

Refuel valve position indicator light does not come on with valves selected open and refuel- ing manifold pressurized.

4）电门在关闭位时，保持亮。

Stay on with switch at CLOSED.

9. 加油。

Refueling.

1）禁止进行所有的加油。

All refueling is prevented.

2）通风油箱燃油溢出。

Fuel spills at surge tank.

3）电门在关闭位时活门不关闭。

Valve does not close with switch at CLOSED.

Task: Liu Shuai finds out some problems in the check and has some writing tasks. The first one is that the light does not show when valves selected are open and refueling manifold is pressurized, the second is that light does not go off after pressing the press-to-test switch, and the last is to do the operational test of fuel quantity indicating system. Please finish the writing tasks for him.

Warning Case

2016 Chapecoenese Crash Caused by Being Lack of Fuel

In 2016, a Brazilian football team was destroyed by the Chapecoenese air crash. Seventy-one of 77 passengers on-board died on November 28, including 19 Chapecoenese players and all the club's coaching staff.

12 警示案例——沙佩科人空难：燃油不足

Warning Tips: the captain and first officer have found the problem of insufficient fuel, but during the flight, the airline did not land and refuel midway in accordance with aviation regulations in order to reduce expenses, which is also the direct cause of the plane crash.

Lesson 4   Hydraulic System

Learning Objectives

1. Knowledge objectives:
   A. To grasp the words, related terms and abbreviations about hydraulic system.
   B. To grasp the key sentences about hydraulic system.
   C. To know the main components of hydraulic system.

2. Competence objectives:
   A. To be able to read and understand frequently-used & complex sentence patterns, capitalized English materials and obtain key information on hydraulic system quickly.
   B. To be able to communicate with English speakers about hydraulic system.
   C. To be able to fill in job cards in English.

3. Quality objectives:
   A. To be able to self-study with the help of aviation dictionaries, the Internet and other resources.
   B. To develop the craftsman spirit of carefulness and responsibility.

Section 1   Aviation Listening & Speaking

1.1 Aviation Listening: listen to the record and fill in the blanks.

1 听力录音   2 答案及原文

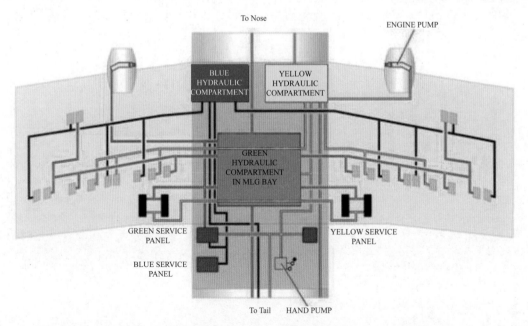

What is that barking sound coming from below the cabin floor as soon as we reach the gate position? Have you ever wondered what is going on in the cargo compartment where this very strange noise is coming from. Many people refer to this as the "Airbus Barking Dog Sound".

This strange barking sound emits from the PTU, the 1._____Transfer Unit which is a substantial part of the Airbus hydraulic system. I explain in further 2._____when the PTU is being operated and it's purpose.

Now the Airbus A320 3._____consist of three independent system, named in colours, green, blue and yellow and one of their highest priorities is to pressure the actuators to move the flight controls. The 4._____ system is powered by an engine driven hydraulic pump, engine No. 1, so in flying direction the engine to your left hand side and the yellow system is powered by engine No. 2 but can also be powered by an electric backup pump. The blue system is powered by an electric 5._____only but isn't associated with the PTU. It does, however, have its own unique backup mechanism, the Ram Air Turbine, but that's a whole another video.

So imagine the PTU as a reversible motor pump located between the green and yellow systems. Hydraulic fluid from either system can 6._____the pump to 7._____the other hydraulic system. But the fluids only drive the pump so they remain isolated from each other, meaning they don't mix.

Each hydraulic system normally 8. _____at 3000 PSI and if the difference between the green and yellow system 9._____500 PSI, the system 10._____the PTU to pressurize the low system. In other words, the green or yellow hydraulic pump has failed and is being pressurized from the other system through the PTU.

1.2 Aviation Speaking: practice the following dialogue and design your own.

Situation:

This is a dialogue between the pilot and the ground crew. In the process of pushing the aircraft, the crew may apply for hydraulic pressure or start the engine. Practice it and have your own.

Note: PIL=Pilot, GND=Ground Crew, CTL= Air Traffic Controller, CCA=Air China.

3 对话录音

Scene 1

GND: Cockpit, Ground.

PLT: Go ahead.

GND: Ground ready.

PLT: Confirm clear to pressurize the hydraulics?

GND: Clear to pressurize the hydraulics.

PLT: Confirm clear to start engine No. 1 and No.2.

GND: Clear to start both engines.

PLT: Engines start completed. Please disconnect.

GND: Disconnect headset. See you on your right/left side. Bye-bye.

4 对话译文

Scene 2

PIL: Confirm clear to pressurize the hydraulics?

GND: Clear to pressurize the hydraulics.

PIL: Confirm clear to start engine No. 1?

GND: Clear to start engine No. 1.

Words & Expressions

1. confirm [kənˈfɜːm] *v.*   确认
2. disconnect [ˌdɪskəˈnekt] *v.*   断开
3. headset [ˈhedset] *n.*   头戴式耳机话筒

Section 2 Aviation Reading

5 课文朗读   6 课文译文

Pre-reading questions:

1. Is there a hydraulic system on an aircraft?
2. What's the function of hydraulic system?

7 飞机上的液压系统（微课）

Hydraulic System

Introduction

The main function of the hydraulic system is to provide power control for the operation of primary and secondary flight controls, landing gear, brakes, nose wheel steering and thrust reversers. The Boeing 757 had three independent hydraulic systems. For identification, they are color-coded: red for the left system, blue for the center system and green for the right system.

Components Description

A basic hydraulic system consists of a reservoir, an engine driven pump, a filter, a pressure regulator, an accumulator, a pressure gage, a relief valve, an emergency pump, two check valves, a selector valve and an actuating cylinder. Here only some of the components are briefly introduced.

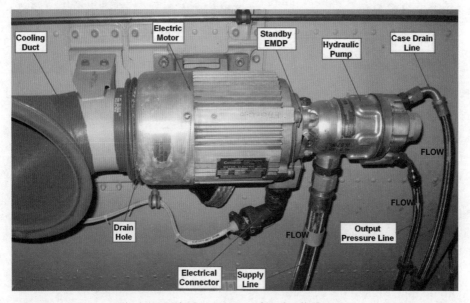

Fig.4-1   Main Components of Hydraulic System

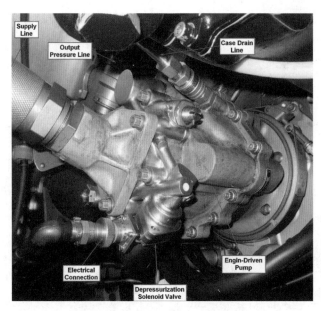

Fig.4-2   Engine-Driven Pump

The hydraulic reservoir is a tank or a container to store sufficient hydraulic fluid for all conditions of operation. It replenishes the system fluid when needed and provides room for thermal expansion.

Hydraulic pumps are designed to cause fluid-flow. Hydraulic power is transmitted by movement of fluid by a pump. The pump does not create the pressure, but the pressure is produced when the flow of fluid is restricted.

A filter is a screening or straining device used to clean the hydraulic fluid, thus preventing foreign particles and contaminating substances from remaining in the system. If such objectionable material is not removed, it may cause the entire hydraulic system of the aircraft to fail through the breakdown or malfunctioning of a single unit of the system.

Hydraulic pressure must be regulated in order to use it to perform the desired tasks.

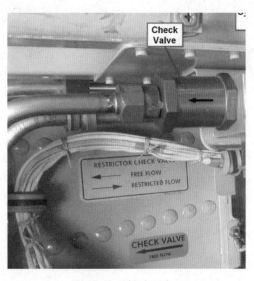

Fig.4-3   Check Valve

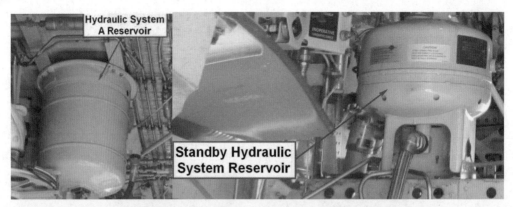

Fig.4-4   Hydraulic System A Reservoir & Standby Hydraulic System Reservoir

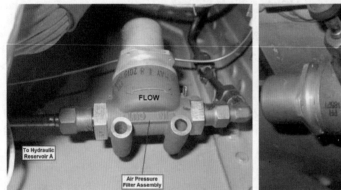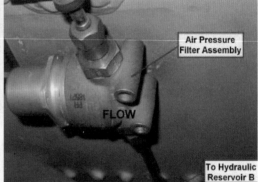

Fig.4-5   Air Pressure Filter Assembly

Fig.4-6   Air Pressure Gauge   Fig.4-7   Air Pressure Relief Valve

There are three independent hydraulic systems that supply hydraulic power for user systems to power the flight controls, the landing gear, and thrust reversers. System A and system B are full-time operating system during flight that share responsibility for all hydraulically powered components. The standby system is operated only on demand. The main and auxiliary hydraulic systems supply pressurized fluid to both thrust reversers, power transfer unit (PTU) motor, landing gear extension and retraction, nose wheel steering, main gear brakes, primary flight controls, and secondary flight controls.

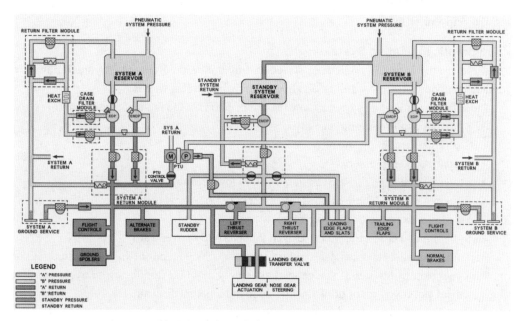

Fig.4-8   Schematic for B737-700 Hydraulic

All Hydraulic Systems on an Airplane

The hydraulic power system consists of these sub-systems of main hydraulic systems, ground servicing system, auxiliary hydraulic systems and hydraulic indicating systems.

Main Hydraulic Systems

The main hydraulic systems are A and B. System A has most of its components on the left side of the airplane and system B on the right side. The Boeing 737-1/200 had system A powered by the two Engine Driven Pumps (EDP's) and system B powered by the two Electric Motor Driven Pumps (EMDP's). There is also a ground interconnect switch to allow system A to be powered when the engines are shut down. From the Boeing 737-300 onwards each hydraulic system had both an EDP and an EMDP for greater redundancy in the event of an engine or generator failure.

The EDP's are much more powerful, having a hydraulic flow rate of 22 gpm (Classics)/37 gpm (NG). The EMDP's only produce 6 gpm. The standby system output is even less at 3 gpm.

Note that the EDP's do not have an OVERHEAT light. This is because they are mechanically (not electrically) driven and have very little heat rise so there is no need for an overheat warning. Note also that the EDP's are always working when the engine is turning, so they can not be disconnected or switched off. Switching an EDP off leaves the pump running but opens a pressure relief bypass valve to take the fluid away from the pump.

Ground Servicing System

The ground servicing system fills all hydraulic reservoirs from one central location.

Auxiliary Hydraulic Systems

The auxiliary hydraulic systems are the standby hydraulic system and the power transfer unit (PTU) system. The standby hydraulic system is a demand system that supplies reserve hydraulic power to these components of rudder, leading edge flaps and slats, and both thrust reversers. The hydraulic PTU system is an alternative source of hydraulic power for the leading edge flaps and slats, and auto-slat system.

Hydraulic Indicating Systems

These are the hydraulic indicating systems of hydraulic fluid quantity, hydraulic pressure, hydraulic pump low pressure warning, and hydraulic fluid overheat warning.

The hydraulic indicating systems show these indications in the flight compartment:

➢ System A and B reservoir quantity.

➢ Standby reservoir low quantity.

➢ System A and B pressure.

➢ System A and B Engine-Driven Pump (EDP) low pressure.

➢ System A and B Electric Motor-Driven Pump (EMDP) low pressure.

➢ Standby electric motor-driven pump low pressure.

➢ System A and B Electric Motor-Driven Pump (EMDP) overheat.

Warning: hydraulic fluid can cause injury to persons. If you get the hydraulic fluid on your skin, flush your skin with water. If you get hydraulic fluid in your eyes, flush your eyes with water and get medical aid. If you eat or drink the hydraulic fluid, get medical aid.

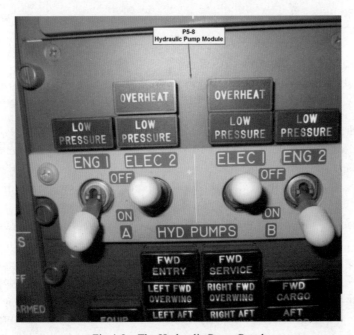

Fig.4-9   The Hydraulic Pump Panel

Pressurization

Air pressure from the reservoir pressurization system maintains head pressure on hydraulic system A, B, and the standby hydraulic system reservoirs. The pressurized reservoirs supply a constant flow of fluid to the hydraulic pumps.

Hydraulic system A

Hydraulic system A supplies pressure to these airplane systems:

➢ Power transfer unit motor.

➢ Left thrust reverser.

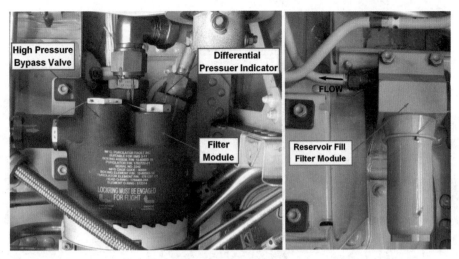

Fig.4-10   Reservoir Filter Module

➢ Landing gear extension and retraction.
➢ Nose wheel steering.
➢ Alternate brakes.
➢ Aileron.
➢ Autopilot A.
➢ Elevators.
➢ Elevator feel.
➢ Flight spoilers 2, 4, 9, and 11.
➢ Ground spoilers 1, 6, 7, and 12.
➢ Rudder.

Hydraulic system B

Hydraulic system B supplies pressure to these airplane systems:
➢ Right thrust reverser.
➢ Alternate landing gear retraction.
➢ Alternate nose wheel steering.
➢ Normal brakes.
➢ Aileron.
➢ Autopilot B.
➢ Elevators.
➢ Elevator feel.
➢ Flight spoilers 3, 5, 8, and 10.
➢ Rudder.
➢ Trailing edge flaps.
➢ Leading edge flaps and slats.

Standby Hydraulic System

The standby hydraulic system supplies alternative hydraulic pressure to these airplane systems of thrust reversers, standby rudder, and leading edge flaps and slats.

Hydraulic PTU System

Fig.4-11   Hydraulic PTU System

The hydraulic PTU system is a hydraulic motor-pump assembly that supplies alternative pressure to leading edge flaps and slats and auto slat system if system B is depressurized. The PTU is controlled by the PTU control valve. System A pressurizes the motor when the PTU control valve is open. System B supplies the fluid to the pump.

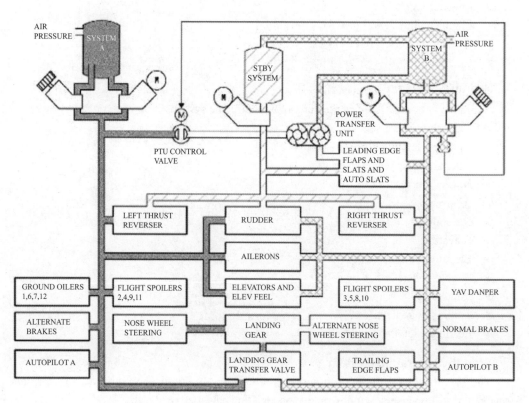

Fig.4-12   Hydraulic Power

The Operation of Hydraulic System

Pressurize the hydraulic systems with a ground service cart or the hydraulic pumps. The switches on the hydraulic panel control hydraulic system A and B hydraulic pumps. The flight control panel lets you control the standby hydraulic system.

Ground Service Cart Pressurization

To pressurize hydraulic system A, connect a ground service cart to the left ground service disconnect. To pressurize hydraulic system B, connect a ground service cart to the right one. The standby system can not be pressurized from a ground service cart.

Hydraulic Pump Pressurization

Use the hydraulic panel to turn on and monitor the hydraulic pumps for system A and B. The flight control panel lets you turn on the standby pump. Pressurize system A and B with either an EDP or an EMDP. The ELEC 1 and ELEC 2 switches on the hydraulic panel control the EMDPs. When the hydraulic pressure is normal, the hydraulic low pressure lights will go off. Normally, the EDPs are on. When the engines are on, the EDPs come on to also pressurize systems A and B. An overheat light monitors the system temperature.

If the hydraulic systems are pressurized with the hydraulic pumps, make sure there is sufficient fuel in the main fuel tanks to cool the heat exchangers. If they are pressurized with a ground service cart, the pressure must be removed from the hydraulic reservoir first.

When hydraulic power is supplied, keep persons and equipment away from all control surfaces (the ailerons, elevators, rudder, flaps, slats and spoilers) and the nose gear, for they are supplied with power by the hydraulic system. Injury to persons or damage to equipment can occur when hydraulic power is supplied.

To make sure the hydraulic systems operate correctly, monitor the instruments and indicator lights when they are pressurized. If the overheat light comes on, the operation must be stopped immediately, or damage to the equipment can occur. If a main fuel tank is less than 250 gallons, do not operate the EMDP for more than two minutes. Before the pump is operated again, let the reservoir temperature decrease to ambient temperature or damage to equipment can occur.

Abbreviations & Acronyms

1. Engine-Driven Pump (EDP)　　　　　发动机驱动泵
2. Electric Motor-Driven Pump (EMDP)　电动马达驱动泵
3. Electrical (ELEC)　　　　　　　　　电的
4. Power Transfer Unit (PTU)　　　　　动力转换装置

8 单词朗读

Words & Expressions

1. hydraulic [haɪˈdrɔlɪk] *adj.*　　　　　液压的
2. primary [ˈpraɪˈmeri] *adj.*　　　　　基本的
3. secondary [ˈsɛkəndɛri] *adj.*　　　　辅助的
4. thrust [θrʌst] *n.*　　　　　　　　　推力
5. thrust reverser　　　　　　　　　　反推装置
6. identification [aɪˌdɛntəfɪˈkeʃən] *n.*　辨认
7. reservoir [ˈrɛzəˌvɔr] *n.*　　　　　（液压）油箱
8. driven [ˈdrɪvn] *adj.*　　　　　　　驱动的
9. filter [ˈfɪltə] *n.*　　　　　　　　　过滤器；油滤

9 术语专讲：液压
（微课）

10. pressure regulator   调压器

11. accumulator [ə'kjumjəletə] *n.*   蓄压器

12. relief valve   释压活门；安全阀

13. selector valve   选择活门；选择阀

14. cylinder ['sɪlɪndə] *n.*   气缸

15. replenish [rɪ'plɛnɪʃ] *v.*   加油

16. thermal expansion   热膨胀

17. screening ['skrinɪŋ] *v.*   屏蔽，筛选

18. foreign ['fɔrən] *adj.*   外来的

19. particle ['pɑrtɪkl] *n.*   粒子

20. foreign particle   杂质粒子

21. objectionable [əb'dʒekʃ(ə)nəb(ə)l] *adj.*   反对的，有异议

22. breakdown ['brek'daʊn] *n.*   故障

23. malfunctioning[mæl'fʌŋkʃəniŋ] *n.*   失效，故障

24. motor ['motə] *n.*   电机，马达

25. nose wheel steering   前轮转弯

26. autopilot ['ɔtəʊpaɪlət] *n.*   自动驾驶仪

Exercises

2.1 Choose the best answer from A, B, C, D options.

10 答案

1. actuating cylinder

    A. 内筒   B. 作动筒   C. 外筒   D. 减震筒

2. nose wheel steering（N/WS）

    A. 前轮转弯   B. 前轮舱   C. 机身轮子转弯   D. 转弯作动筒

3. ground interconnect

    A. 中性点接地   B. 互联系统   C. 互联电力   D. 接地互联

4. overheat(OVHT)

    A. 头顶   B. 过热   C. 大修   D. 超速

5. thermal expansion

    A. 热吸收   B. 突然膨胀   C. 超声膨胀   D. 热膨胀

6. 制动

    A. break   B. branch   C. brake   D. bake

7. 反推装置

    A. thrust reverser   B. reverse thrust   C. gross thrust   D. reverse thrust nozzle

8. 发动机驱动泵

    A. Integrated Driven Generator(IDG)   B. Engine Driven Pump (EDP)

    C. Air Driven Generator (ADG)   D. Electric Motor Driven Pump(EMDP)

9. 备用泵

    A. standby pump   B. injection pump   C. boost pump   D. emergency pump

10. 马达

    A. generator   B. regulator   C. actuator   D. motor

2.2 Find the best option to paraphrase the sentence.

1. The standby system is operated only on demand.

    A. There is no need to operate the standby system.

B. Operate the backup system only when needed.

C. The standby system can be operated at any time.

D. The main system is operated only on demand.

2. The ground servicing system fills all hydraulic reservoirs from one central location.

A. All hydraulic reservoirs are filled up through the servicing system from all directions.

B. All hydraulic reservoirs are drained through the servicing system from one central location.

C. From a central location, the ground servicing system can make all hydraulic oil tanks full.

D. The ground servicing system fills all hydraulic reservoirs from many ways.

3. The flight control panel lets you control the standby hydraulic system.

A. You can use the flight control panel to control the standby hydraulic system.

B. You can't control the standby hydraulic system through the flight control panel.

C. The flight control panel lets you control the main hydraulic system.

D. The standby hydraulic system can't be operated through the flight control panel.

4. An overheat light monitors the system temperature.

A. The system temperature is monitored by an overheat light.

B. The runway light monitors the system temperature.

C. An overheat light monitors the system pressure.

D. An overhead light supervises the system temperature.

5. When hydraulic power is supplied, keep persons and equipment away from all control surfaces.

A. When hydraulic power is supplied, keep persons and equipment close to all control surfaces.

B. All persons and equipment should be kept away from all control surfaces.

C. When the engine is running, keep people and equipment away from the engine.

D. When hydraulic power is provided, persons and equipment should be kept away from all control panels.

2.3 Change the following components specified in lower case and translate them into Chinese.

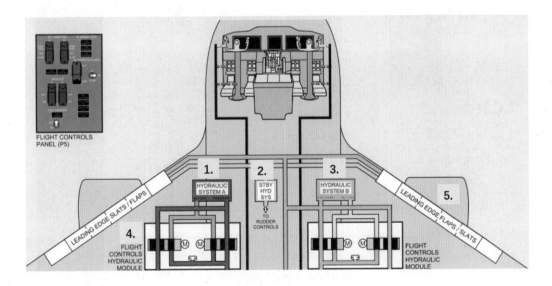

1.	
2.	
3.	
4.	
5.	

Section 3 Aviation Translation: translate the following sentences into Chinese

11 空军 "金牌蓝天工匠" 施娟：成功的人舍得真心付出

1. NOTE: IF THE PRESSURE DECREASES, THERE MAY BE AN AIR LEAK IN THE RESERVOIR PRESSURIZATION SYSTEM.

2. WARNING: HYDRAULIC FLUID CAN CAUSE INJURIES TO PERSONS. IF YOU GET THE HYDRAULIC FLUID ON YOUR SKIN, FLUSH YOUR SKIN WITH WATER. IF YOU GET HYDRAULIC FLUID IN YOUR EYES, FLUSH YOUR EYES WITH WATER AND GET MEDICAL AID. IF YOU EAT OR DRINK THE HYDRAULIC FLUID, GET MEDICAL AID.

Section 4 Aviation Writing

Liu Shuai is an aircraft mechanic. He is inspecting the plane and finds some faults.

Please help him write down fault descriptions. Some related words, phrases & terms and key sentences are offered as follows.

Key words, phrases & terms:

1. 液压泵   hydraulic pump

2. 液压接通   hydraulic power on

3. 液压管路   hydraulic line

4. 液压作动筒   hydraulic actuator

5. 压力传感器   pressure sensor

6. 热交换器   heat exchanger

7. 密封圈   seal

8. 储油箱   reservoir

9. 排放油滤   drain filter

10.释压活门   relief valve

11.流量控制活门   flow control valve

12. 压力控制活门   pressure control valve

13. 液压过热警告电门   hydraulic overheat warning switch

14. 刹车压力指示   brake press indicator

15. 管路   plumbing

16. 冲压涡轮   RAT

Key sentences:

1. 从液压油箱通气管道流出。

   Comes out of the reservoir vent line.

2. 从液压油箱通气管道排放杆流出。

   Comes out of the reservoir vent line drain mast.

3. 从 A 系统传输到 B 系统。

   Transfers from system A to system B.

4. 航后检查 A 系统回油滤组件内丢失一弹簧。

   Do AF check and find one spring of A system return filter ASSY lost.

5. 航后检查发现右发吊架上部有一根液压系统回油管接头漏油，检查为管接头破裂。

   A fitting of the hydraulic system's return line in the R/H is found leaking in AF check.

   The fitting has a crack after inspection.

6. 液压储压器的勤务。

   Serve hydraulic accumulator.

7. 航后检查发现 B 系统液压管路 T 型接头漏油。

   The T-type union of the hydraulic system B leaks oil in AF check.

8. 更换液压地面加油滤。

   Replace hydraulic ground service filter.

9. 更换正释压活门滤滤芯。

   Replace SAF checkout relief valve filter elements.

10. 更换 A 系统液压系统回油滤。

    Replace System A hydraulic return filter.

Task: Liu Shuai has three writing tasks. The first one is that there is fluid leakage, the second is to replace the seal, and the last one is to serve hydraulic reservoirs. Please finish the writing tasks for him.

12 警示案例——联合
航空 232 号班机
苏城空难

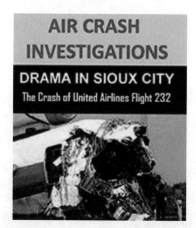

AIR CRASH
INVESTIGATIONS

DRAMA IN SIOUX CITY
The Crash of United Airlines Flight 232

Warning Case

Sioux Crash of United Airline 232

The United Airlines Flight 232 accident occurred in Sioux, Iowa, USA on July 19, 1989 due to machine failure, resulting in 110 deaths among 285 passengers and 1 death among 11 crew members. As the accident occurred in Sucheng, it is also known as "Sucheng Air Crash".

Warning Tips: the Douglas DC-10 airliner flying this flight was damaged by the fragments emitted after the blade was broken due to the material problem of the tail engine fan, resulting in the loss of control of the aircraft. The cause of the accident belongs to mechanical failure.

Lesson 5 Pneumatic System

Learning Objectives

1. Knowledge objectives:

    A. To grasp the words, related terms and abbreviations about pneumatic system.

    B. To grasp the key sentences about pneumatic system.

    C. To know the main components of pneumatic system.

2. Competence objectives:

    A. To be able to read and understand frequently-used & complex sentence patterns, capitalized English materials and obtain key information quickly.

    B. To be able to communicate with English speakers about the topic freely.

    C. To be able to fill in the job cards in English.

3. Quality objectives:

    A. To be able to self-study with the help of aviation dictionaries, the Internet and other resources.

    B. To develop the craftsman spirit of carefulness and responsibility.

Section 1 Aviation Listening & Speaking

1.1 Aviation Listening: listen to the record and fill in the blanks.

1 听力录音   2 答案及原文

In this lesson, we will discuss the high volume, low 1._____air systems used by modern jet airliners. The system is normally supplied with 2._____air from the engine compressors. This can be backed up with air from the auxiliary 3. _____air supply carts. This air is used for air conditioning and 4. _____Unit (APU) on the ground by external ice protection, and as a power source for the air turbine motors used for engine starting driving hydraulic pumps and air driven 5._____and slats. It is also used for pressurizing hydraulic reservoirs, potable water tanks, and cargo compartment 6._____. The engine bleed air system consists of the power source, the engine compressor, and control devices for temperature and pressure regulation during 7._____, because of the great variation of our 8._____ pressure from a gas turbine engine between idle a maximum RPM. There is a need to maintain a reasonable 9._____ of air during lower RPM as well as restricting excessive pressure when the engine is at high RPM. This is usually done by taking air from two 10._____ stages to maintain a reasonable pressure band at all engine speeds.

1.2 Aviation Speaking: practice the following dialogue and design your own.

Situation: This is a dialogue between the pilot in the cockpit and the mechanic on the ground. On the field a plane is ready to be pushed back. Practice it and have your own.

Note: PIL=Pilot, GND=Ground Mechanic

3 对话录音

4 对话译文

Scene

PIL: How is the ground checks, please?

GND: Ground checks completed, preparing to push the airplane.

PIL: Parking brake is released.

GND: Commence pushing back.

PIL: Push-back completed, please set the parking brake.

GND: Cleared to start engines.

PIL: Affirmative, cleared for engines starting.

GND: All engines started, cleared to disconnect interphone and show me the disconnect interphone and the pin, goodbye.

PIL: Understand and all engines started. Please standby for the pin and hand signal on your left, goodbye.

Words & Expressions

1. commence [kə'mens] *v.*　　　　　开始
2. signal ['sɪgnəl] *n.*　　　　　　信号

Section 2 Aviation Reading

5 课文朗读　　6 课文译文

Pre-reading questions:

1. What does the pneumatic system control?
2. What does the pneumatic system provide?

Pneumatic System

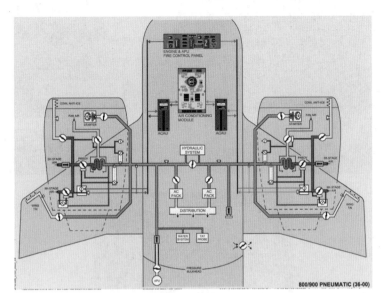

7 气动系统
（微课）

Fig.5-1　Schematic for B737-800/900 Pneumatic System

Introduction

The purpose of the pneumatic system is to supply compressed air for a controlled temperature and pressure environment during all phases of flight and ground operation. Air is obtained from en- gine bleeds, APU, or an external ground cart through a connector located on the fuselage. The system controls the temperature and the pressure of engine bleed air source. It provides high temperature and high pressure air for engine starting, cabin pressurization and air conditioning, wing anti-ice, water reservoir pressurization and hydraulic reservoir pressurization on some aircraft. The engine bleed air systems are located on the engine and within the support strut. You may find other systems and components that also work pneumatically, for instance, air-driven pumps for hydraulics or air motors used for slat operation.

Operating Principles

The manifold is normally split by the isolation valve. With the isolation valve switching in AUTO, the isolation valve will only open when an engine bleed air or pack switch is selected OFF.

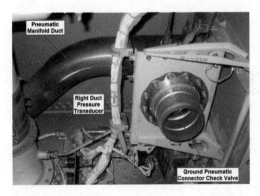

Fig.5-2　Ground Pneumatic Connector Check Valve

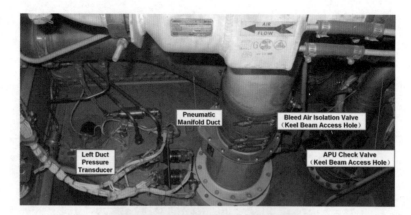

Fig.5-3   Pneumatic Manifold

Air for engine starting, air conditioning packs, wing anti-ice and the hydraulic reservoirs comes from their respective ducts. Air for pressurization of the water tank and the aspirated TAT probe comes from the left pneumatic duct. External air for engine starting feeds into the right pneumatic duct. Ground conditioned air feeds directly into the mix manifold.

The minimum pneumatic duct pressure (with anti-ice off) for normal operation is 18 PSI. On the Max, the pneumatic bleed air system now has an electronic controller. This allows the aircraft to digitally tune the amount of air that is needed in whatever flight regime you're in. This is different to the previous "all or nothing" system which would often take more bleed air from the engines than necessary thereby reducing performance.

Fig.5-4   TAT Probe (Aspirated-1)

Fig.5-5   TAT Probe (Unaspirated-2)

Different Air Sources

The sources of pneumatic power are as follows:

➢ Engine 1 bleed air system.

➢ Engine 2 bleed air system.

➢ Auxiliary Power Unit (APU) bleed air system.

➢ Pneumatic ground air connection.

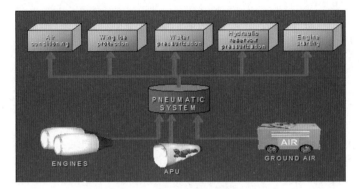

Fig.5-6　Sources of Pneumatic Power

The pneumatic manifold collects the compressed air from the sources and supplies it to the user systems. In any pneumatic system, air pressure and temperature are needed to be as constant as possible to support the different consumers effectively. On all jet aircraft there is a very good air source available. The engine compressor provides enough air for combustion purpose and also the pneumatic system. You can see that the compressed air is bled from the engines, so it is called Engine Bleed Air. The engines are not the only source of air supply for the pneumatic system. APU is used to supply the pneumatic system. The only built-in air pressure sources are the engines and the APU. If the aircraft is on ground, the pneumatic system can be supplied with external air. The third air source can be delivered via connectors and the High Pressure Ground Connectors.

Components Location

The hydraulic reservoirs are pressurized to ensure a positive flow of fluid, which reaches Pump A from the left manifold and B from the right. See wheel-well FWD.

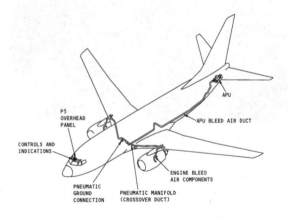

Fig.5-7　Components Locations of Pneumatic System

The engine bleed system components are on the engine compressor cases and in the engine struts. The pneumatic manifold (crossover duct) is in these areas:

➢ The leading edges of the wing inboard of the engine struts.

➢ The forward areas of the air conditioning pack bays.

➢ The keel beam.

APU and the APU bleed system components are in a torque box in Section 48. The APU bleed air duct is in these areas:

➢ The right side of the APU torque box.

➢ Section 48 and the aft pressure bulkhead.

➢ The left side and forward bulkhead of the aft cargo compartment.

➢ The keel beam.

The pneumatic system controls and indications are on the air conditioning/bleed air controls panel in the flight compartment.

The Systems Using Pneumatic Power

The following systems use pneumatic power:

➢ Engine start systems.

➢ Air conditioning and pressurization systems.

➢ Engine inlet cowl anti-ice systems.

➢ Wing thermal anti-ice systems.

➢ Water tank pressurization system.

The engine bleed air systems control the temperature and the pressure of engine bleed air source and are located on the engine and within the support strut. The purpose of the pneumatic system is to supply bleed air from the 5th-stage and 9th-stage of the engine compressor, APU, or a ground cart. Air is distributed by a pneumatic manifold from the above sources to the air conditioning packs, wing and cowl Thermal Anti-Ice (TAI) systems, the engine starting system, potable water system and the hydraulic reservoir.

Fig.5-8　A Propeller with an Electro-thermal Deice System

Pneumatic Lights and Indications

If engine bleed air temperature or pressure exceeds limits, the BLEED TRIP OFF light will illuminate and the bleed valve will close. You may use the TRIP RESET switch after a short cooling period. If the BLEED TRIP OFF light does not extinguish, it may be due to an overpressure condition. Bleed trip off's are most common on full thrust, bleeds off, take-off's. The reason is excessive leakage past the closed hi-stage valve butterfly which leads to a pressure build-up at the downstream port on the overpressure switch within the hi-stage regulator. The simple in-flight fix is to reduce duct pressure by selecting CLB-2 and/or using engine and/or wing anti-ice.

Fig.5-9   737-400 Pneumatics Panel & 737-800/900 Pneumatics Panel

WING-BODY OVERHEAT indicates a leak in the corresponding bleed air duct. This is particularly serious if the leak is in the left hand side, as this includes the ducting to APU. The wing-body overheat circuits may be tested by pressing the OVHT TEST switch; both wing-body overheat lights should illuminate after a minimum of 5 seconds. This test is part of the daily inspection.

Fig.5-10   Cabin Logbook

Pneumatic Distribution Systems

Pneumatic distribution system supplies compressed air to the user systems. The distribution system has these subsystems:

> Engine bleed air system.
> APU bleed air system.
> Pneumatic ground air connection.
> Pneumatic manifold system.

There are two engine bleed air systems, one for each engine. Engine bleed air is from the 5th and 9th stages of the engine high pressure compressors. At low engine speeds, bleed air comes from the 9th stage. At high engine speeds, bleed air comes from the 5th stage.

The APU load compressor supplies bleed air on the ground and in the air.

Fig.5-11   Engine Bleed Air System

The pneumatic ground air connection provides for the connection of a ground pneumatic cart. The pneumatic cart can supply bleed air for engine start and ground use of the air conditioning system.

The pneumatic manifold system gets bleed air from the engines, APU, or pneumatic ground cart. A bleed air isolation valve divides the pneumatic manifold into left and right sides. The normal position of this valve is closed. Because of this, a single duct failure cannot cause a loss of pressure to the whole pneumatic manifold. The pneumatic manifold has two pressure transmitters. One transmitter is on the right side and the other is on the left side. They measure the manifold pressures. Pressure indication is on the air conditioning panel. The right side of the manifold has a pneumatic ground air connector. Valves on the pneumatic manifold control the flow of air to the user systems. The user systems control the operation of these valves.

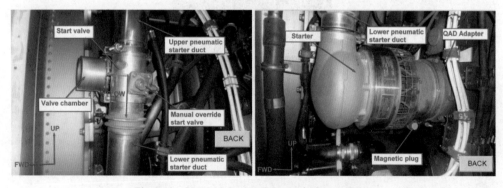

Fig.5-12   Upper & Lower Pneumatic Starter Duct

Warnings & Cautions

CAVTION: THE AIR IN THE PNEUMATIC SYSTEM IS HOT AND UNDER HIGH PRESSURE. MAKE SURE TO DEPRESSURIZE THE PNEUMATIC SYSTEM BEFORE WORKING ON IT.

CAUTION: ALWAYS APPLY ELECTRICAL POWER BEFORE APPLYING PNEUMATIC POWER, AND REMOVE PNEUMATIC POWER BEFORE REMOVING ELECTRICAL

POWER TO PREVENT POSSIBLE DAMAGES TO THE EQUIPMENT.

CAUTION: WHEN WORKING IN THE FACILITY OF ELECTRICAL STATIC SENSITIVE DEVICES, A GROUNDING WRIST -STRAP MUST BE USED.

WARNING: PROVIDING PNEUMATIC POWER WILL SUPPLY PRESSURE TO THE OPERATING SYSTEMS. CARE SHOULD BE TAKEN TO ISOLATE THOSE SYSTEMS AND CONTROLS NOT INTENDED FOR OPERATION TO PREVENT LOSS OF PRESSURE TO PREVENT INADVERTENT ACTUATION OF THE EQUIPMENT, DAMAGES TO AIRPLANE AND INJURIES TO PERSONNEL.

WARNING: DO NOT SUPPLY MORE THAN 60 PSI OF PRESSURE TO THE PNEUMATIC SYSTEM. IF YOU SUPPLY TOO MUCH PRESSURE, DAMAGES TO EQUIPMENT AND INJURIES TO PERSONNEL CAN OCCUR.

Abbreviations & Acronyms

1. Isolation Valve(ISO VALVE)　　　　隔离活门
2. Overheat (OVHT)　　　　　　　　过热
3. Pounds per Square Inch(PSI)　　　　磅／平方英寸
4. Total Air Temperature(TAT)　　　　空气总温
5. Thermal Anti-ice(TAI)　　　　　　热空气防冰
6. Transmitter (XMTR)　　　　　　　传感器

8 单词朗读

Words & Expressions

1. pneumatic [njuːˈmætɪk] *adj.*　　气动的
2. compress [kəmˈpres] *v.*　　　　压缩
3. engine bleed　　　　　　　　　发动机引气
4. cabin [ˈkæbɪn] *n.*　　　　　　机舱，客舱
5. hydraulic [haɪˈdrɒlɪk] *adj.*　　液压的
6. pressurization [ˌpreʃəraɪˈzeɪʃn] *n.*　增压
7. air-driven *adj.*　　　　　　　空气驱动的
8. slat [slæt] *n.*　　　　　　　前缘缝翼
9. duct [dʌkt] *n.*　　　　　　　管道
10. aspirated TAT probe　　　　　吸气式总温探头
11. manifold [ˈmænɪfəʊld] *n.*　　总管，歧管
12. wheel-well　　　　　　　　　轮舱
13. compressor [kəmˈpresə(r)] *n.*　压气机
14. flight regime　　　　　　　　飞行状态
15. ground connector　　　　　　地面连接器
16. fluid [ˈfluːɪd] *n.*　　　　　液体，流体
17. crossover duct　　　　　　　交输管道
18. bay [beɪ] *n.*　　　　　　　舱，隔舱
19. keel beam　　　　　　　　　龙骨梁
20. torque box　　　　　　　　　扭矩盒
21. bulkhead [ˈbʌlkhed] *n.*　　　隔板
22. cowl [kaʊl] *n.*　　　　　　整流罩
23. ground cart　　　　　　　　地面气源车
24. illuminate[ɪˈluːmɪneɪt] *v.*　　照亮

9 术语专讲：舱（微课）

25. trip reset	跳闸复位
26. regulator ['regjuleɪtə(r)] *n.*	调节器
27. bleed valve	引气阀，放气阀
28. extinguish [ɪk'stɪŋgwɪʃ] *v.*	熄灭
29. overpressure [ˌəʊvə'preʃə] *n.*	超压
30. leak [liːk] *n.&v.*	泄漏
31. ground pneumatic cart	地面气源车
32. depressurize [diːˈpreʃəraɪz] *v.*	使……减压
33. static sensitive device	静电敏感元件
34. wrist-strap	防静电手环

10 答案

Exercises

2.1 Choose the best answer from A, B, C, D options.

1. pneumatic
   A. 液压   B. 气动   C. 电动   D. 电子
2. connector
   A. 联系   B. 人脉   C. 含义   D. 接头
3. air driven pump(ADP)
   A. 气动泵   B. 风动发电机   C. 发动机驱动泵   D. 气压表
4. jet aircraft
   A. 民用飞机   B. 军用飞机   C. 喷气式飞机   D. 旋翼飞机
5. probe
   A. 问题   B. 过程   C. 特性   D. 探头
6. 压缩空气
   A. compressed air   B. heavy hydrogen   C. vacuum bottle   D. compressed oxygen
7. 总温
   A. ATC   B. TAT   C. FCC   D. TCAS
8. 平方英寸
   A. cubic meter   B. cubic foot   C. square inch   D. square foot
9. 超压
   A. low pressure   B. high pressure   C. differential pressure   D. overpressure
10. 交叉管道
   A. bypass   B. crossover duct   C. cross road   D. crossover frequency

2.2 Find the best option to paraphrase the sentence.

1. The system controls the temperature and the pressure of engine bleed air source.
   A. The system controls the temperature and the friction of engine bleed air source.
   B. The temperature and the pressure of engine bleed air source are operated by the system.
   C. Another system controls the temperature and the pressure of engine bleed air source.
   D. The system only controls the temperature of engine bleed air source.
2. The engine compressor provides enough air for combustion purpose and the pneumatic system.
   A. The engine compressor doesn't provide enough air for combustion purpose and the pneumatic system.
   B. The engine compressor only can provide enough air for combustion purpose.

C. Enough air for combustion purpose and the pneumatic system is supplied by the engine compressor.

D. The engine combustor can't provide enough for combustion purpose and the pneumatic system.

3. If the aircraft is on ground, the pneumatic system can be supplied with external air.

A. If the aircraft is in the air, external air cannot supply air to the pneumatic system.

B. If the aircraft is on ground, the hydraulic system can be supplied with external air.

C. If the aircraft is in the air, external air can supply air to the compressor.

D. If the aircraft is on the ground, external air can supply air to the pneumatic system.

4. A bleed air isolation valve divides the pneumatic manifold into left and right sides.

A. The pneumatic manifold is divided into left and right sides by a bleed air isolation valve.

B. A bleed air isolation valve is divided into left and right sides by the pneumatic manifold.

C. A bleed air isolation valve divides the hydraulic manifold into left and right sides.

D. A bleed air isolation valve divides the hydraulic manifold into three parts.

5. The air in the pneumatic system is hot and under high pressure.

A. The air in the hydraulic system is hot and under high pressure.

B. The air in the pneumatic system is hot and under low pressure.

C. The air in the pneumatic system is cool and under low pressure.

D. The air in the pneumatic system is under high temperature and high pressure.

2.3　Change the following components specified in lower case and translate them into Chinese.

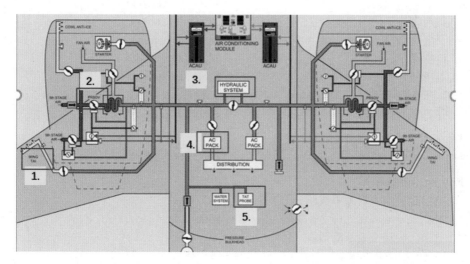

1.	
2.	
3.	
4.	
5.	

Section 3 Aviation Translation: translate the following sentences into Chinese

11 忘填工卡日志（文本）

1. WARNING: DO NOT SUPPLY MORE THAN 60 PSI OF PRESSURE TO THE PNEUMATIC SYSTEM. IF YOU SUPPLY TOO MUCH PRESSURE, DAMAGES TO EQUIPMENT AND INJURIES TO PERSONNEL CAN OCCUR.

2. WARNING : PROVIDING PNEUMATIC POWER WILL SUPPLY PRESSURE TO THE OPERATING SYSTEMS. CARE SHOULD BE TAKEN TO ISOLATE THOSE SYSTEMS AND CONTROLS NOT INTENDED FOR OPERATION TO PREVENT LOSS OF PRESSURE TO PREVENT INADVERTENT ACTUATION OF THE EQUIPMENT, DAMAGES TO AIRPLANE AND INJURIES TO PERSONNEL.

Section 4 Aviation Writing

Liu Shuai is an aircraft mechanic. He is inspecting the plane and finds some faults.

Please help him write down fault descriptions. Some related words, phrases & terms and key sentences are offered as follows.

Key words, phrases & terms:

1. 地面气源   ground pneumatic
2. 超温电门   over temperature switch
3. 单向活门   check valve
4. 引气活门   bleed air valve
5. APU 引气活门   APU bleed air valve
6. 引气隔离活门   bleed air isolation valve
7. 预冷器控制活门传感器   precooler control valve sensor
8. 引气调节器   bleed air regulator
9. 高压调节器   high stage regulator
10. 高压级活门   high stage valve
11. 双压力指示器   dual pressure indicator
12. 总管压力传感器   manifold press transmitter
13. 引气跳开灯   bleed trip light
14. 引气压力   pneumatic pressure

Key sentences:

1. 机组反映空中右发引气跳开灯亮。
   The crew reported that the bleed trip lights of the R/H engine were on in air.
2. 更换压力调节关断活门。
   Replace the PRSOV.
3. 拆除并安装右发高压调节器。
   Remove and install R/H engine high stage regulator.
4. 慢车时左发低气压排故。
   Resolve the troubleshooting of L/H engine low pneumatically pressure at low idle.

Task: Liu Shuai has three writing tasks. The first one is to have a troubleshooting of right engine low pneumatically pressure at low idle, the second is to finish removal and installation of left engine low stage valve, and the last one is to finish the task of bleed air valve. Please finish the writing tasks for him.

Warning Case

MU5759-Malfunction of Bleed Air System

12 警示案例——
MU5759 排气
系统故障

On September 20th, 2015, during the descent of flight MU5759 from Kunming to Shenzhen, the bleed air system showed a malfunction. In order to prevent the cabin decompression, the captain manually released passenger oxygen masks according to flight procedures. The plane landed safely and normally at Shenzhen Airport, and no passengers were injured.

Warning Tips: The cabin of civil aircraft has a special air conditioning system, which pressurizes the cabin and controls air flow, air filtration, air temperature and humidity. Once this system fails or the body is damaged in the air, it is necessary to put down the oxygen mask to meet the basic breathing needs of passengers.

Lesson 6 Water and Waste System

Learning Objectives

1. Knowledge objectives:

   A. To grasp the words, related terms and abbreviations about water and waste system.

   B. To grasp the key sentences about water and waste system.

   C. To know the main components of water and waste system.

2. Competence objectives:

   A. To be able to read and understand frequently-used & complex sentence patterns, capitalized English materials and obtain key information about water and waste system.

   B. To be able to communicate with English speakers freely.

   C. To be able to fill in job cards in English.

3. Quality objectives:

   A. To be able to self-study with the help of aviation dictionaries, the Internet and other resources.

   B. To develop the craftsman spirit of carefulness and responsibility.

Section 1 Aviation Listening & Speaking

1.1 Aviation Listening: listen to the record and fill in the blanks.

1 听力录音   2 答案及原文

Water pressures vary in different locations of a 1._____system. Water mains 2._____the street may operate at higher 3._____, with a pressure 4._____ located at each point where the water enters a building or a house. In poorly managed 5._____, water pressure can be so low as to result only in a trickle of water or so high that it leads to damage to 6._____fixtures and waste of water. Pressure in an 7._____water system is typically maintained either by a pressurized water 8._____serving an urban area, by pumping the water up into a water 9._____and relying on gravity to maintain a 10._____pressure in the system or solely by pumps at the water treatment plants and repeater pumping stations.

1.2 Aviation Speaking: practice the following dialogue and design your own.

Situation:

This is a dialogue between the pilot and the ground mechanic about pushing back the aircraft. After boarding, the ground mechanic contacts the crew to set the brakes. Practice it and have your own.

Note: PIL=Pilot, GND=Ground Mechanic

3 对话录音

Scene 1

GND: Cockpit, Ground.

PIL:　Go ahead.

GND: Set parking brakes.

PIL:　Parking brakes set.

GND: Roger.

Scene 2

GND: Cockpit, Ground.

PIL:　Go ahead.

GND: Ready for push-back. Release the parking brakes.

PIL:　Parking brakes released. You can push-back to Tango 2(center line), face east.

GND: Push-back to Tango 2, face east.

Scene 3

GND: Push-back to remote stand No.43.

PIL:　Remote stand No.43.

4 对话译文

Words & Expressions

remote stand　　　　　　　　远机位

Section 2　Aviation Reading

Pre-reading questions:

1. Which do you prefer on the aircraft, water or drink?

2. Do you know where the waste goes?

5 课文朗读　　6 课文译文

Potable Water and Waste System

7 水 & 废水系统
（微课）

Fig.6-1　Water Tanker

Potable water and waste system supplies potable water to the lavatories and galleys and removes sink and toilet waste. It also removes rain water from the door sill areas. The potable water system supplies potable water to the galleys and lavatories while the waste system is for storing and removing waste from the toilets. This system is composed of independent but related systems.

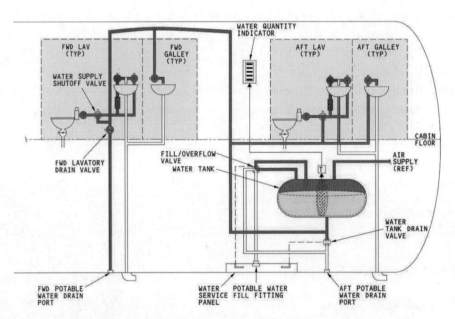

Fig.6-2　Schematic for Potable Water & Water System

There is a 60 US Gal tank (227 litres) behind the aft cargo hold for potable water. This serves the galleys and washbasins, but not the toilets as they use chemicals. Waste water is either drained into the toilet tanks or expelled through heated drain masts. The tank indicator (Boeing 737-300/-400/-500) is located over the rear service door. Press to test; indications are clockwise from 7 o'clock: Empty, 1/4, 1/2, 3/4, Full. The NG has an LED panel that is always lit (below) for both potable water and waste tank.

The water and waste system includes the following subsystems of potable water, air supply and waste disposal. The potable water system supplies water to the lavatories and galleys. The air supply system pressurizes the water tank. The waste disposal system removes waste from the lavatories, galleys and door sill drains.

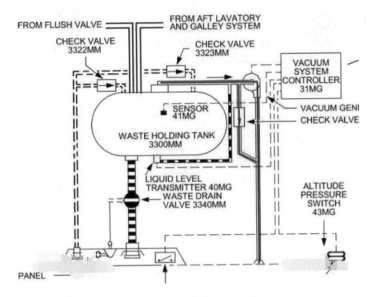

Fig.6-3　Schematic for Waste System

Fig.6-4　Galley & Lavatory

Potable Water System

The potable water system consists of a water tank, quantity indication, pressurization components and distribution tubing.

Fig.6-5　Water Tank

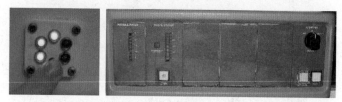

Fig.6-6　Potable Water Quantity Indicator–Classics & NG

　　The water tank and pressurization components are located aft of the aft cargo compartment. The quantity indication is in the aft section of the passenger cabin. The water tank stores potable water for use by the passengers and crew. The tank is constructed of fiberglass and attached by struts and mounting brackets to the airplane structure. The 30 U.S. gallon tank is cylindrical. It is protected from freezing by a 3-piece fiberglass blanket. On the upper section of the tank are connections for the air pressure line, fill line, overflow line, supply line and a quantity transmitter, the fill and overflow valve is used to fill the tank to the capacity determined by a standpipe, this valve is operated by a handle. A drain valve is located on the bottom of the tank.

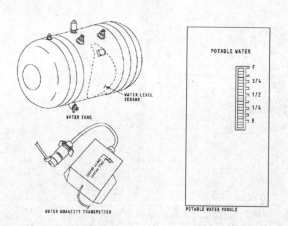

Fig.6-7　Water Quantity Transmitter & Water Quantity Indication

　　The quantity indication shows the amount of water in the tank. The quantity transmitter consists of 10 reed switches and 3 inside tube. A float with 3 magnets surrounds the tube. In addition, the indicator consists of 5 lights. The tank determines the position of the float. The magnets on the float close the associated reed switches. Use the push button on the quantity indicator. It allows the corresponding quantity lights to illuminate. If the tank is in emptiness, the E light is illuminated and if the tank is full, all lights are illuminated.

The potable water system has the subsystems of passenger water, water heating and water quantity indication. The passenger water system supplies water to the lavatories and galleys. The water heating system heats the water supplied to the lavatory hot water faucets. The water quantity indication system measures and displays the quantity of water in the potable water system. The potable water system has one tank that holds potable water.

Fill and drain the water tank at the water service panel. To fill the water tank, open the fill/overflow valve and add water through the potable water fill fitting until the water flows from the potable water drain port.

The water tank supplies water to galley faucets, lavatory faucets and lavatory toilets.

Each lavatory has a water supply shutoff valve. The water supply shutoff valve lets you isolate the water supply to the sink or to the toilet. The water supply shutoff valve also lets you isolate the sink and the toilet at the same time.

The forward lavatory has a drain valve. The forward lavatory drain valve drains water from the forward water supply lines.

A water heater in each lavatory increases the temperature of the water supplied to the hot water faucets.

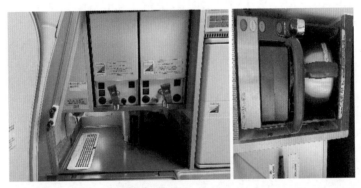

Fig.6-8　Water in the Galley

The water tank level sensor sends the water quantity data to a water quantity transmitter. The water quantity transmitter sends the data to a water quantity indicator at the attendant panel. The water quantity indicator shows the level of water in the tank.

To drain the potable water system, open the valves of water tank drain valve, forward lavatory drain valve (forward lavatory only) and lavatory water supply shutoff valves (one in each lavatory).

When draining the potable water system, the water drains overboard through the forward and aft potable water drain ports.

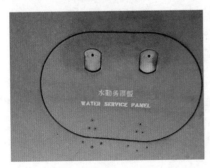

Fig.6-9　Water Service Panel

Fig.6-10　Potable Water Tank Drain Overflow

Fig.6-11　Toilet Service Panel

Warnings & Cautions

CAUTION: WHEN DRAINING THE POTABLE WATER SYSTEM, DRAIN WATER FROM WATER SERVICE PANEL AND FORWARD LAVATORY. THE WATER SERVICE PANEL LETS YOU DRAIN THE WATER TANK AND AFT SUPPLY LINES. THE FORWARD LAVATORY HAS A DRAIN VALVE LETTING YOU DRAIN THE FORWARD WATER SUPPLY LINES. WHEN DRAINING THE POTABLE WATER SYSTEM, TURN THE WATER SUPPLY SHUTOFF VALVES IN EACH LAVATORY TO THE SUPPLY ON POSITION. IF YOU DO NOT TURN THE WATER SUPPLY SHUTOFF VALVE TO THE SUPPLY ON POSITION, WATER WILL NOT DRAIN FROM THE TOILET OR LAVATORY FAUCETS.

WARNING: IF THE POTABLE WATER SYSTEM IS NOT DRAINED AT A MINIMUM OF ONE TIME EACH THREE DAYS, THE GROWTH OF BACTERIA CAN OCCUR. IF BACTERIA GROWTH CONTINUES, AND YOU DRINK THE WATER, ILLNESS CAN OCCUR.

CAUTION: DRAIN THE WATER SYSTEM. IF THE WATERLINES HAVE WATER IN THEM, THEY CAN FREEZE IN COLD WEATHER. THIS CAN CAUSE DAMAGES TO THE WATER LINES.

CAUTION: YOU MUST FULLY DRAIN THE POTABLE WATER SYSTEM BEFORE YOU ADD A DISINFECTANT OR WHEN YOU PARK THE AIRPLANE IN COLD WEATHER.

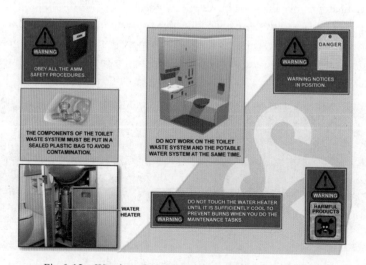

Fig.6-12　Warnings & Cautions for Potable Water & Waste

WARNING: SEAL THE CARGO COMPARTMENT WITH THE LINING. OBEY THE INSTRUCTIONS IN THE SPECIFIED PROCEDURE WHEN YOU INSTALL THE LINING. IF YOU INSTALL THE LINING INCORRECTLY, SMOKE CAN GET INTO THE PASSENGER COMPARTMENT DURING A FIRE.

CAUTION: REMOVE THE CARGO COMPARTMENT LINING TO GET ACCESS TO SOME OF THE POTABLE WATER SYSTEM COMPONENTS. MAKE SURE THAT REPLACE THE LINING CORRECTLY WHEN YOU ARE DONE.

Waste Disposal System

Waste disposal system removes water from the lavatory and galley sinks, human waste from the lavatory toilets, and rain water from the door sills. It has the subsystems of gray water, vacuum waste and waste tank quantity indication.

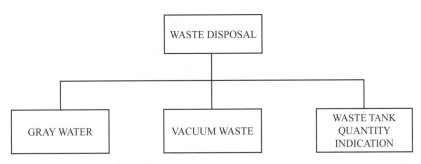

Fig.6-13　Subsystems of Waste Disposal

The gray waste comes from the lavatory and galley sinks, and from door sill drains at each entry and galley service door. This system drains water and other liquids from the lavatory and galley sinks, gives exhaust ventilation from the lavatories and galleys, and drains rain water from the entry/service door sill areas. The gray water from the sinks drains overboard through drain masts. The gray water from the door sills drains overboard through drain fittings.

The vacuum waste system removes human waste from the toilets. The waste material is held in a waste tank until servicing. The waste tank quantity indication system measures and shows the level of waste in the waste tank.

Fig.6-14　Waste Disposal

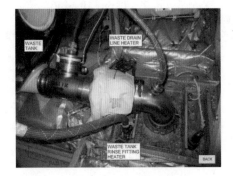

Fig.6-15　Waste Tank

Abbreviations & Acronyms

1. Empty (E) 空的

2. Heater (HTR)   加热器，预热器

3. Indication (IND)   指示

4. Lavatory (LAV)   厕所

5. Light Emitting Diode (LED)   发光二极管

6. Quantity (QTY)   数量，量

Words & Expressions

8 单词朗读

9 术语专讲：真空（微课）

1. potable water   饮用水

2. disposal [dɪ'spəʊzl] *n.*   处理，清理

3. lavatory['lævətri] *n.*   厕所

4. galley ['gæli] *n.*   厨房

5. door sill *n.*   门框

6. drain mast *n.*   排放杆

7. subsystem [səb'sɪstəm] *n.*   子系统

8. air supply system   供气系统

9. water quantity indication   水量指示

10. cylindrical [sə'lɪndrɪkl] *adj.*   圆柱形的

11. construct [kən'strʌkt] *v.*   制作

12. bracket ['brækɪt] *n.*   支架

13. fiberglass['faɪbəglɑːs] *n.*   玻璃纤维

14. blanket ['blæŋkɪt] *n.*   充注管道

15. transmitter [trænz'mɪtə(r)] *n.*   传送器

16. overflow line   外泄管道

17. capacity [kə'pæsəti] *n.*   容积

18. standpipe['stændpaɪp] *n.*   立管

19. drain [dreɪn] *n.*   排水；下水道

20. tank [tæŋk] *n.*   水箱，油箱

21. reed [riːd] *n.*   簧片

22. magnet['mægnət] *n.*   磁铁

23. float [fləʊt] *n.*   浮子

24. associate [ə'səʊsieɪt] *adj.*   关联的

25. corresponding[ˌkɒrə'spɒndɪŋ] *adj.*   相应的

26. faucet['fɔːsɪt] *n.*   水龙头

27. water service panel   水服务面板

28. galley faucet   厨房水龙头

29. lavatory toilet   厕所马桶

30. water supply shutoff valve   供水关闭阀

31. water tank level sensor   水箱液位传感器

32. waste quantity indicator   废物量指标

33. water tank drain valve   水箱排水阀

34. bacteria[bæk'tɪəriə] *n.*   细菌

35. water line   水管

36. disinfectant[ˌdɪsɪn'fektənt] *n.*   消毒剂

37. lining['laɪnɪŋ] *n.*   内衬，内装饰板

38. human waste   人类排泄物

39. vacuum waste system 　　　　　真空废物系统
40. gray water 　　　　　　　　　　灰水
41. drain fitting 　　　　　　　　　排放接头

10 答案

Exercises

2.1 Choose the best answer from A, B, C, D options.

1. potable water
   A. 纯净水　　　　B. 饮用水　　　　C. 矿物质水　　　　D. 便携水
2. lavatory
   A. 洗脸盆　　　　B. 下水道　　　　C. 厕所　　　　　　D. 厨房
3. galley
   A. 厨房　　　　　B. 画廊　　　　　C. 走廊　　　　　　D. 过道
4. door sill
   A. 门框　　　　　B. 门廊　　　　　C. 门槛　　　　　　D. 门禁
5. drain mast
   A. 溢流口　　　　B. 排放系统　　　C. 电缆杆　　　　　D. 排放杆
6. LED
   A. 发光二极管　　B. 液晶显示屏　　C. 前缘　　　　　　D. 后缘
7. 加仑
   A. liter　　　　　B. gram　　　　　C. gallon　　　　　D. pound
8. 升
   A. meter　　　　　B. liter　　　　　C. litter　　　　　D. little
9. 立管
   A. test stand　　　B. engine stand　　C. force pipe　　　D. standpipe
10. 灰水
   A. potable water　B. grey water　　C. mineral water　D. purified water

2.2 Find the best option to paraphrase the sentence.

1. The water and waste system is composed of independent but related systems.
   A. The water and waste system consists of independent but related systems.
   B. The water and waste system is composed of independent but irrelated systems.
   C. The water and waste system is composed of dependent but related systems.
   D. The disposal system is composed of independent but related systems.

2. The waste disposal system removes waste from the lavatories, galleys, and door sill drains.
   A. The distribution system removes waste from the lavatories, galleys, and door sill drains.
   B. Waste from the lavatories, galleys, and door sill drains is discharged by the waste disposal system
   C. The drainage system deals with waste from the lavatories, galleys, and door sill drains.
   D. The waste disposal system contains waste from the lavatories, galleys, and door sill drains.

3. The quantity indication shows the amount of water in the tank.
   A. The quality indication shows the amount of water in the tank.
   B. The quantity indication shows the rest of water in the tank.

C. The quantity indication shows the waste in the tank.

D. The quantity indication shows the quantity of water in the tank.

4. The water heating system heats the water supplied to the lavatory hot water faucets.

A. The water heating system heats the water supplied to the galley hot water faucets.

B. The water heating system heats the water supplied to the lavatory cold water faucets.

C. Water heated by the water heating system is supplied to the lavatory hot water faucets.

D. Water heated by the water heating system is supplied to the galley hot water faucets.

5. The vacuum waste system removes human waste from the toilets.

A. The vacuum waste system disposes of human waste from the toilets.

B. The vacuum waste system removes waste from the galley.

C. The vacuum waste system removes gray water from the lavatory.

D. The vacuum waste system contains human waste from the toilets.

2.3 Change the following components specified in lower case and translate them into Chinese.

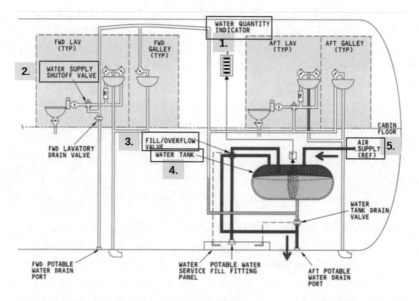

1.	
2.	
3.	
4.	
5.	

Section 3 Aviation Translation: translate the following sentences into Chinese

NOTE: THE DISINFECTANT MUST STAY IN THE POTABLE WATER SYSTEM FOR ONE HOUR MINIMUM.

11 航修焊接专家孙红梅——
一颗匠心护"战鹰"（文本）

WARNING: IF THE POTABLE WATER SYSTEM IS NOT DRAINED AT A MINIMUM OF ONE TIME PER THREE DAYS, THE GROWTH OF BACTERIA CAN OCCUR. IF BACTERIA GROWTH CONTINUES, AND YOU DRINK THE WATER, ILLNESS CAN OCCUR.

Section 4 Aviation Writing

Liu Shuai is an aircraft mechanic. He is inspecting the plane and finds some faults.

Please help him write down fault descriptions. Some related words, phrases & terms and key sentences are offered as follows.

Key words, phrases & terms:

1. 水 / 污水箱   water/ waste tank
2. 排水口   drain mast
3. 加注 / 溢流活门   fill/overflow valve
4. 球形活门   ball valve
5. 排放管 / 接头   drain line/ fitting
6. 饮用水箱压缩机   water tank compressor
7. 水量指示器   water quantity indicator
8. 水勤务面板   water service panel
9. 控制手柄   control handle
10. 空气过滤器   air filter
11. 废物量指标   waste quantity indicator
12. 水压限制电门   water pressure limit switch
13. 不工作 / 测试电门   INOP/TEST switch
14. 厕所水加热器   lavatory water heater
15. 厕所卷筒纸架   lavatory roll holder
16. 盥洗室供水   water supply in lavatory

Key sentences:

1. 前舱 102 号热水器不工作，没有热水。

   NO.102 water heater in FWD galley does not work and no hot water supplied.

2. 前舱 203 号热水器在空中及地面一直显示 NO WATER 打开水阀后，右边凉水管一直漏水。

   No. 203 water heater in FWD galley always shows "NO WATER" both in flight and on ground, and the right cold water pipe always leaks after turning on the water valve.

3. 前舱服务间 102 号热水器关不紧，总滴水。空中只有关闭总阀。

   The valve of 102 water heater in forward galley could not be entirely shut off and the supply pipe is always closed in flight.

4. 前厨房的烧水杯不加热。

   The hot cup at forward galley can not operate normally.

5. R2 门洗手间盆池水龙头水流不止。

   The water tap could not be stopped in R2 lavatory.

6. R2 门处洗手间洗手池无水，水阀已打开。

   The closet of the lavatory R2 has no water and the water valve has been open.

7. 更换厨房的水滤滤芯。

   Replace the water filter element of galleys.

Task: Liu Shuai has three writing tasks. The first one is that the cold faucet in forward galley leaks, the second is to remove and install water tank drain valve, and the last one is to serve potable water system. Please finish the writing tasks for him.

Warning Case

Air Canada 797- Fire Fight

12 警示案例——
加拿大航空

On June 2nd , 1983, A DC-9-32 airliner traveled from Dallas, USA to Montreal, Canada. The aircraft's sensor indicated that the aircraft's lavatory was blocked by a foreign object, and the circuit breaker tripped. The captain tried to re-open the breaker but was unsuccessful. A fire broke out in the lavatory, and thick smoke filled the plane. The fire after landing made it too late for the 23 passengers on board to escape and they all died.

Warning Tips: First, the escape routes and facilities should be improved. Second, smoke detectors are mandatory on all civil aircraft.

Lesson 7 Engine Oil

Learning Objectives

1. Knowledge objectives:

   A. To grasp the words, related terms and abbreviations about engine oil.

   B. To grasp the key sentences about engine oil.

   C. To know the function and main components of engine oil.

2. Competence objectives:

   A. To be able to read and understand frequently-used & complex sentence patterns, capitalized English materials and obtain key information on engine oil.

   B. To be able to communicate with English speakers freely.

   C. To be able to fill in job cards in English.

3. Quality objectives:

   A. To be able to self-study with the help of aviation dictionaries, the Internet and other resources.

   B. To develop the craftsman spirit of carefulness and responsibility.

Section 1 Aviation Listening & Speaking

1.1 Aviation Listening: listen to the record and fill in the blanks.

1 听力录音   2 答案及原文

In this lesson, we will examine the properties of the various fuels used in aircraft engines. Aircraft piston engines use gasoline and gas 1._____engines, kerosene. Both of these fuels are produced from crude 2._____. In simple terms, the required fuel is extracted from the crude oil by distillation process. The oil is boiled in a furnace, and then the oil vapor is 3._____in a distillation column. The various hydrocarbons in oil all have different 4._____points, so they condense out of the vapor and are tapped off at different points in the column. The heavy oils 5._____ out first and are collected near the bottom. The process then continues up through the tower with kerosene and gasoline, which are relatively light and volatile, being taken out near the top. The specification from ideal fuel for either a gas turbine engine or a piston engine include the following main requirements. It should flow easily under all 6._____conditions. It should have complete 7._____ under all conditions. It needs to have a high calorific value. The calorific value is a measure of the amount of heat released during its combustion. It should be non corrosive. There should be no 8._____ to the engine from combustion by products. The fuel should present a low fire hazard. Engines should start easily, and the fuel must be able to lubricate the moving parts of the fuel 9. _____ and other 10._____ in the system. In practice, the cost of satisfying all these criteria is prohibitive.

1.2 Aviation Speaking: practice the following dialogue and design your own.

3 对话录音

Situation:

This is a dialogue between the pilot and the ground mechanic. The pilot reports there is a not a sufficient quantity of oil in the tank. The ground mechanic is going to fill the oil tank. Practice it and have your own.

Note: PIL=Pilot, GND=Ground Mechanic

Scene 1

GND: The crew member reports the engine oil is not sufficient.

PLT: Have you checked it in the cockpit?

GND: Yes. The quantity of oil is below 2.5 U.S. gallons (9.5 liters).

PLT: Have you checked it through the sight gage'?

GND: Yes, there is also a bright view through the sight gage. The indication shows that there is not a sufficient quantity of oil in the tank for airplane dispatch.

PLT: We must fill the oil tank. Preparate to fill the oil tank.

GND: Roger.

PLT: Don't forget to clean the oil tank scrapper with a cloth.

GND: Roger.

4 对话译文

Scene 2

GND: There are fuel fumes when I remove oil filler cap. And there is fuel in the oil tank, I think we must replace main oil / fuel heat exchanger and the servo fuel heater.

PLT: Roger. Flush the engine oil system if it contains contamination.

Words & Expressions

1. gage [geɪdʒ] *n.*　　　　　　　测量仪表
2. dispatch [dɪˈspætʃ] *n.*　　　　派遣
3. scrapper [ˈskræpə] *n.*　　　　刮板
4. fume [fjuːm] *n.*　　　　　　　烟
5. oil filler cap　　　　　　　　　加油口盖
6. servo [ˈsɜːvəʊ] *n.*　　　　　　伺服

7. contamination [kən,tæmɪ'neɪʃn] *n.*　　污染

Section 2 Aviation Reading

Pre-reading questions:
1. Why is oil needed?
2. Where is it used?

5 课文朗读　　6 课文译文　　7 发动机滑油（微课）

Engine Oil

Generally, lubrication is used to reduce friction between metal surfaces that move against each other. The contact surfaces look very smooth but when you look more closely at them through a microscope, you can see that they are very rough. When the surfaces move against each other, they can cause very high friction and wear. So Engine oil is needed to form a protective oil film. This prevents the contact between the metal surfaces. An oil pump pumps the oil into the gearbox and then it is squeezed into the gap between two parts. These tasks of oil include lubricating, cooling, cleaning, and corrosion protection.

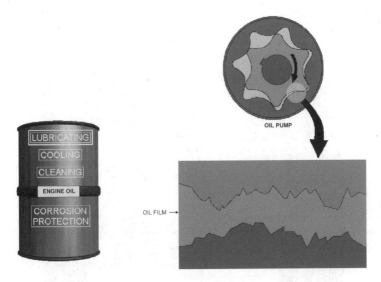

Fig.7-1　Functions of Engine Oil

The oil system is pressurized by the engine driven oil pump. The oil leaves the oil pump, where sensors for oil pressure and the LOW OIL PRESSURE switch are located, passes through an oil filter and continues to the engine bearing and gearbox, and it returns to the oil tank by engine driven scavenge pump. From the scavenge pumps the oil passes through a scavenge filter. Bypass valve is mounted to prevent filter from blocking. Scavenge oil temperature is sensed as the oil returns to the scavenge pump. The oil then passes through the servo fuel heater and fuel/oil heat exchanger where it is cooled by fuel before returning to the oil tank.

The main components include the oil tank, the supply lines, the supply pump and the supply filter. The oil tank is on the fan case, at the 3:00 position. Do the oil level check and fill the oil tank through the oil tank access door. The oil tank access door is on the side of the right fan cowl. You can also open the right fan cowl to get access to the oil tank.

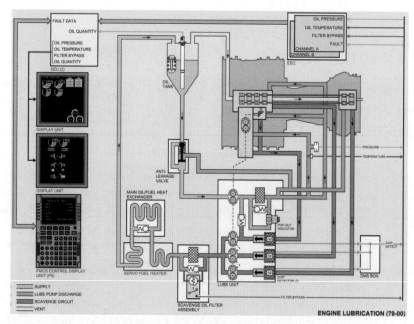

Fig.7-2　Schematic for Engine Oil

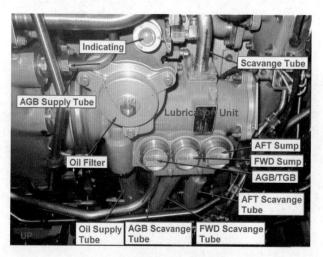

Fig.7-3　Engine Oil Components -1

Fig.7-4　Engine Oil Components -2

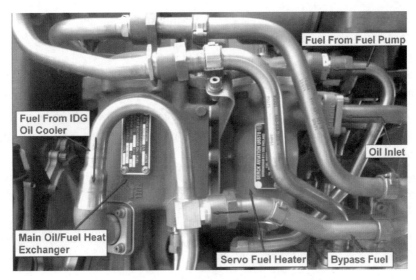

Fig.7-5   Engine Oil Components -3

The oil is stored in the oil tank and pumped by the supply pump through the supply lines to the oil nozzles in the engine bearing compartments and the gearboxes. The oil is filtered before it reaches the oil nozzles. The supply filter is located downstream of the supply pump. The filter removes any foreign particles from the oil before it reaches the oil nozzles. This prevents blockage of the nozzles.

The oil reservoir and supply system are also called the pressure oil system.

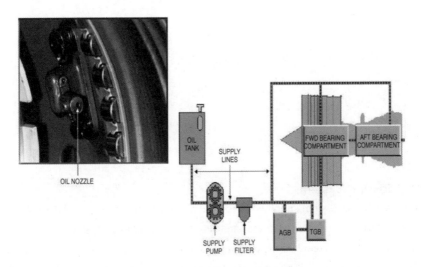

Fig.7-6   Pressure Oil System

Subsystems

The engine oil system supplies oil to lubricate, cool, and clean the engine bearings and gears. It has these subsystems of storage, distribution and indicating.

Oil Storage System

The oil storage system keeps sufficient oil for a continuous supply to the oil distribution circuit. The oil is kept in the engine oil tank. The functions of engine oil tank include containing the engine oil, removing the air from the scavenge oil, doing an oil level check and filling the oil system.

The oil tank has an oil level sight gage, a gravity fill port, and pressure servicing fill ports. Use the oil level sight gage on the oil tank to make a visual check of the quantity. The oil level sight gage is on the front of the oil tank. Use the oil tank gravity on the right of the oil tank fill port to fill the oil tank. The oil filler cap has a locking handle. The oil that falls during servicing collects into the oil scupper. The oil scupper connects to a drain line.

A drain plug at the bottom of the oil tank lets you drain it. The oil tank holds approximately 21 US quarts (20.2 liters). The oil tank for engine 2 can hold more oil than engine 1. This is because of the dihedral of the wings.

Do an oil level check and fill the oil tank for the conditions of normal servicing, after replacement of an oil system component and engine oil change.

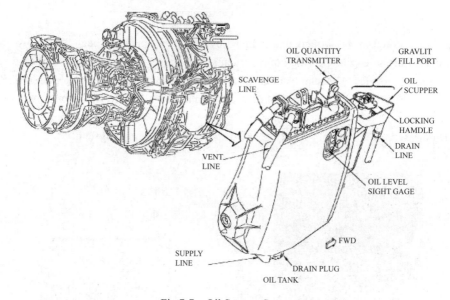

Fig.7-7　Oil Storage System

Oil Distribution System

The oil distribution system supplies oil to cool and lubricate the engine bearings and gears. It also removes oil from sumps and gearboxes and sends it to the storage system.

The oil distribution system has the circuits of supply, scavenge and vent.

The supply circuit sends oil to lubricate the engine bearings and gears. Oil from the oil tank goes to the lubrication unit through an anti-leakage valve. The lubrication unit pressurizes and filters the oil. It then goes to the engine.

The scavenge circuit takes the oil from the engine. The oil first flows through the lubrication unit, which also scavenges the oil. It goes to the scavenge oil filter and then to the servo fuel heater. The oil goes from the servo fuel heater to the main oil/fuel heat exchanger and then back to the servo fuel heater. Then the oil flows back to the oil tank.

The vent circuit balances the internal air pressures in the oil system. Externally, a vent line connects the engine to the oil tank. Unwanted air pressure goes out of the oil tank through the vent line.

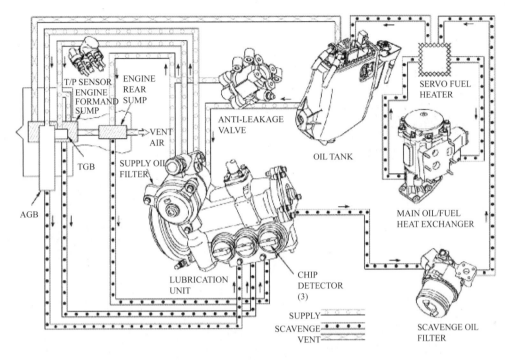

Fig.7-8   Oil Distribution System

Oil Quantity Indicating System

The oil quantity indicating system includes scavenge oil filter bypass indication, low oil pressure indication, oil pressure, oil temperature and oil quantity. This system sends the data to the Display Electronic Units (DEUs). The oil quantity indicating system supplies oil system data to the Display Electronic Units (DEUs).

These components are on the left side of the fan case:

➢ Oil pressure transmitter, on the T/P sensor assembly (10:00 position).

➢ Oil temperature sensor, on the T/P sensor assembly (10:00 position).

➢ Scavenge oil filter clogging transmitter, on the oil scavenge filter assembly (8:00 position).

➢ The oil quantity transmitter, on the oil tank (2:00 position), the right side of the fan case.

The primary and secondary engine displays the data of oil quantity, pressure, temperature and scavenge filter condition on P 2 center instrument panel.

The components including the oil quantity transmitter, oil pressure transmitter, oil temperature sensor, and scavenge oil filter clogging transmitter monitor the oil system.

The oil quantity transmitter sends the oil quantity data directly to the DS/DEUs.

The three other components send data to the DEU through EEC. The temperature/pressure (T/P) sensor assembly contains the oil pressure transmitter and the oil temperature sensor.

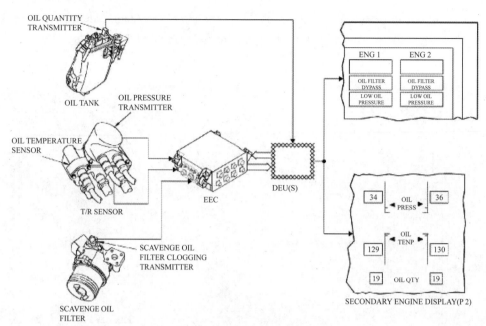

Fig.7-9 Oil Quantity Indicating System

Warnings & Cautions for Engine Oil Servicing

WARNING: DO NOT REMOVE THE FILLER CAP OF THE OIL TANK FOR FIVE MINUTES AFTER AN ENGINE SHUTDOWN. IF THE CHECK VALVE IS DEFECTIVE, HOT OIL CAN SPRAY FROM THE OIL TANK AND CAUSE INJURIES TO PERSONS. THE OIL IS HOT AND PRESSURIZED DURING ENGINE OPERATION.

CAUTION: FULLY CLEAN YOUR SKIN IF YOU TOUCH THE OIL. REMOVE OIL SOAKED CLOTHES IMMEDIATELY. IF THE OIL TOUCHES YOUR SKIN FOR A LONG TIME, IT COULD CAUSE DERMATITIS.

WARNING: DO NOT SERVICE THE OIL TANK WITH OIL BRANDS THAT ARE NOT APPROVED. FLUSH AND REPLACE THE OIL IMMEDIATELY WITH THE CORRECT ENGINE OIL IF BRANDS OF OIL THAT ARE NOT APPROVED ARE USED.

CAUTION: IMMEDIATELY CLEAN THE PAINTED SURFACES ON WHICH THE OIL FALLS. THE OIL WILL PUT STAINS ON CLOTHES AND CAN MAKE SLIGHT PAINT.

CAUTION: ADD THE ENGINE OIL INTO THE GRAVITY FILL PORT UNTIL THE OIL LEVEL GETS TO THE FULL INDICATION ON THE SIGHT GLASS. WHEN THE OIL LEVEL IS AT THE FULL INDICATION ON THE SIGHT GLASS, THE OIL TANK IS FULL.

Abbreviations & Acronyms

1. Common Display System(CDS)　　　　　公共显示系统
2. Display Electronic Units(EDU)　　　　　显示电子组件
3. Electronic Engine Control(EEC)　　　　　发动机电子控制器
4. Temperature/Pressure Sensor(T/P)　　　　温度 / 压力传感器

Words & Expressions

1. lubrication [ˌluːbrɪˈkeɪʃn] *n.*　　　　　　润滑作用

8 单词朗读

9 术语专讲：
火花塞（微课）

2. friction ['frɪkʃ(ə)n] *n.*　　　　摩擦

3. contact surface　　　　接触面

4. microscope ['maɪkrəskəʊp] *n.*　　显微镜

5. metal surface　　　　金属表面

6. rough [rʌf] *adj.*　　　　粗糙的

7. wear [weə(r)] *n.*　　　　磨损

8. protective film　　　　保护膜

9. gearbox ['gɪəbɒks] *n.*　　变速箱；齿轮箱

10. squeeze into　　　　挤入

11. corrosion [kə'rəʊʒ(ə)n] *n.*　　腐蚀

12. filter ['fɪltə(r)] *n.*　　　　过滤器

13. bearing ['beərɪŋ] *n.*　　　　轴承

14. scavenge ['skævɪndʒ] *n.*　　回油

15. servo ['sɜːvəʊ] *n.*　　伺服；伺服系统

16. heater ['hiːtə(r)] *n.*　　　　加热器

17. fan cowl　　　　风扇罩

18. nozzle ['nɒzl] *n.*　　喷嘴；管口

19. foreign particle　　　　杂质颗粒

20. downstream [,daʊn'striːm] *n.*　　下游

21. blockage ['blɒkɪdʒ] *n.*　　堵塞，堵塞

22. reservoir ['rezəvwɑː(r)] *n.*　　油箱

23. storage ['stɔːrɪdʒ] *n.*　　　　存储

24. sufficient [sə'fɪʃ(ə)nt] *adj.*　　足够的

25. scavenge oil　　　　回油

26. sight gage　　　　观测计

27. gravity fill port　　　　重力加油口

28. pressure servicing fill port　　压力加油口

29. continuous [kən'tɪnjuəs] *adj.*　　连续的

30. scupper ['skʌpə(r)] *n.*　　排油口

31. plug [plʌg] *n.*　　插头；塞子

32. quart [kwɔːt] *n.*　　夸脱（容量单位）

33. liter ['liːtə(r)] *n.*　　　　公升

34. dihedral [daɪ'hiːdr(ə)l] *n.*　　二面的，反角

35. vent [vent] *n.*　　出口；通风孔

36. externally [ɪk'stɜːnəli] *adv.*　　外部的

37. pressure transmitter　　　　压力传感器

38. sensor ['sensə(r)] *n.*　　　　传感器

39. clogging ['klɒgɪŋ] *v.& adj.*　　堵塞的

40. secondary ['sekəndri] *adj.& n.*　　辅助的

41. defective [dɪ'fektɪv] *adj.*　　有缺陷的

42. spray [spreɪ] *n.*　　喷雾；喷雾器

43. dermatitis [,dɜːmə'taɪtɪs] *n.*　　皮炎；皮肤炎

44. brand [brænd] *n.*　　标牌，品牌

45. flush [flʌʃ] *n.*　　　　洋溢

46. paint [peɪnt] *n.*　　　　　　油漆
47. sight glass　　　　　　　　观察孔

Exercises

2.1 Choose the best answer from A, B, C, D options.

1. scavenge
   A. 回油　　　　　B. 放油　　　　　C. 加油　　　　　D. 油滤
2. fill port
   A. 加油站　　　　B. 放油　　　　　C. 加注排放　　　D. 排放口
3. gravity fill
   A. 地球重力　　　B. 重力加油　　　C. 万有引力　　　D. 有机填充
4. pressure fill
   A. 重力加油　　　B. 充气　　　　　C. 充液　　　　　D. 压力加油
5. EEC
   A. 欧洲经济共同体　　　　　　　B. 发动机电子控制器
   C. 电子不停车收费　　　　　　　D. 地球地形摄影机
6. 摩擦力
   A. pressure　　　B. friction　　　C. thrust　　　　D. gravity
7. 显微镜
   A. microscope　　B. gyroscope　　C. telescope　　　D. magnifier
8. 滑油
   A. grease　　　　B. fuel　　　　　C. oil　　　　　　D. hydraulic fluid
9. 齿轮箱
   A. torque box　　B. landing gear　C. nose gear　　　D. gearbox
10. 堵塞
    A. blockage　　　B. leakage　　　C. package　　　D. baggage

10答案

2.2 Find the best option to paraphrase the sentence.

1. Generally, lubrication is used to reduce friction between metal surfaces.
   A. Generally, lubrication is used to reduce temperature between metal surfaces.
   B. Generally, hydraulic fluid is used to reduce friction between metal surfaces.
   C. Generally, the friction between metal surfaces is reduced by using lubrication.
   D. Generally, the friction between metal surfaces is increased by using lubrication.
2. The bypass valve is mounted to prevent filter from blocking.
   A. The bypass valve is installed to prevent blockage of filter.
   B. The shutoff valve is mounted to prevent filter from blocking.
   C. The bypass valve is mounted to prevent foreign particles.
   D. The filter prevents blocking.
3. The filter removes any foreign particles from the oil before it reaches the oil nozzles.
   A. The filter keeps any foreign particles in the oil before foreign particles are removed.
   B. The oil nozzles have a component which can filter foreign particles.
   C. The filter is prevented from blocking by removing foreign particles.
   D. The filter disposes of any foreign particles from the oil before it gets into the oil nozzle.
4. A drain plug at the bottom of the oil tank lets you drain it.
   A. A fill plug at the upper of the oil tank lets you fill it.

B. A drain plug located at the bottom of the oil tank to allow you to drain it .

C. There is a drain plug at the bottom of the fuel tank to allow you to fill it.

D. There is a fill plug at the upper of the fuel tank to allow you to fill it.

5. The vent circuit balances the internal air pressures in the oil system.

A. The internal air pressures in the oil system isn't balanced by the vent circuit.

B. The vent circuit keeps the internal air pressure in the oil system balanced.

C. The external air pressures in the oil system is balanced by the vent circuit.

D. The component which balances the external air pressures in the oil system is the vent circuit.

2.3 Change the following components specified in lower case and translate them into Chinese.

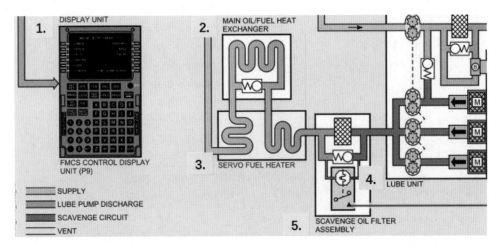

1.	
2.	
3.	
4.	
5.	

Section 3 Aviation Translation: translate the following sentences into Chinese

1. CAUTION: HOLD THE OIL TANK DURING THE REMOVAL OF THE SHOULDER-HEADED PIN. THIS WILL PREVENT DAMAGES TO THE OIL TANK.

2. WARNING:DO NOT REMOVE THE FILLER CAP OF THE OIL TANK FOR FIVE MINUTES AFTER AN ENGINE SHUTDOWN. IF THE CHECK VALVE IS DEFECTIVE, HOT OIL CAN SPRAY FROM THE OIL TANK AND CAUSE INJURIES TO PERSONS. THE OIL IS HOT AND PRESSURIZED DURING ENGINE OPERATION.

11 航修手艺人（文本）

Section 4 Aviation Writing

Liu Shuai is an aircraft mechanic. He is inspecting the plane and finds some faults.

Please help him write down fault descriptions. Some related words, phrases & terms and key sentences are offered as follows.

Key words, phrases & terms:

1. 滑油箱   oil tank
2. 滑油管路   oil line
3. 滑油压力表   oil pressure indicator
4. 滑油温度表   oil temperature indicator
5. 滑油散热器   oil cooler
6. 滑油量传感器   oil quantity transmitter
7. 滑油压力传感器   oil pressure sensor
8. 滑油温度传感器   oil temperature sensor
9. 滑油滤旁通警告电门   oil filter bypass warning switch
10. 润滑剂   lubricant
11. 润滑组件   lubrication unit
12. 回油过滤   scavenge filter
13. 主滑油 / 燃油热交换器   main oil/fuel heat exchanger
14. 燃油 / 滑油冷却器   fuel/cooled oil cooler
15. 发动机滑油消耗高（滑油量快速减少）

high engine oil consumption (oil quantity decreased at a quick rate)
16. 滑油量快速减少   oil quantity decreased at a quick rate

Key sentences:

1. 发动机滑油压力 / 油量指示。

   Engine oil pressure/quantity indication.
2. 断续或空白。

   Intermittent or blank.
3. 低（琥珀色）。

   Low (amber).
4. 低或零（红色）。

   Low or zero (red).
5. 不精确，断续，保持恒定或空白。

   Inaccurate, intermittent, remain constant, or blank.
6. AF 检查发现左发滑油箱口盖金属链丢失。

   Do AF check and find the metal chain of cap to oil tank at L/H is lost.
7. 航后检查发现后货舱滑油储备盒缺少 2 罐滑油。

   Two bottles of oil in the oil store box has been found lack during AF check.
8. 航后检查发现 EEC 自检有 79-31342 滑油压力 DISAGREE 信息。

   The information DISAGREE of the oil pressure appeared when doing the BITE of EEC AF check.
9. 重力添加发动机滑油。

   Add the engine oil by gravity filling procedure.

10. 往启动机加滑油。

Fill the starter with oil.

11. 更换回油滤滤芯。

Replace scavenge oil filter element.

12. 通过重力加油程序给发动机滑油勤务。

Serve the engine oil by gravity fill procedure.

Task: Liu Shuai has three writing tasks. The first one is to fill the oil for constant speed device, the second is to replace oil supply filter element, and the last one is that the oil pressure warning system of the left engine shows (red light) when shutting down the engines. Please finish the writing tasks for him.

Warning Case

Alaska 261-Unlubricated Threads are Worn Bald

On January 31st, 2000, the horizontal stabilizer of Alaska 261 was jammed and fell off during the flight, and the aircraft plummeted into the Pacific Ocean with no survivors. The horizontal stabilizer fell off due to improper maintenance that made the threads went bald without lubricating oil. The pilot tried to solve the problem on his own without guidance and ground support, causing the stuck part to fall off and worsened the situation.

12 警示案例——阿拉斯加航空 261 号班机——无润滑油的螺纹磨秃

Warning Tips: The chief technician reported the loopholes in the company's procedures, and FAA did not thoroughly find out the reasons, nor did it re-supervise the company's complete rectification and maintenance procedures; The management of the company arbitrarily extended the time for maintenance and parts replacement at the expense of maintenance and repair procedures; The maintenance personnel forged work records and did not do the maintenance work that should be done.

Captain Thompson: Center, Alaska 261. we are in a dive here.

HORIZONTAL STABILIZER

Module Two   Electrical and Avionic Systems

Introduction

An aviation maintenance mechanic must have a solid foundation in basic electrical and electronic principles and a good working knowledge of the way applying these principles to complex systems. Electrical systems provide the muscles for retracing landing gears and starting engines and serve as the brains for electronic flight control and monitoring systems.

For many years the only electronics involved in aviation were used for communications and navigation, and all electronic equipment was classified simply as "radio". Today's aircraft employ vast quantities of electronic equipment, much of it unrelated to either communication or navigation.

The term *avionics* is a blend word of aviation and electronics. Avionics are the electronic systems used on the aircraft, artificial satellites and spacecraft. Avionic systems include communications, navigation, the display and management of multiple systems, and the hundreds of systems that are fitted to aircraft to perform individual functions. These can be as simple as a searchlight for a police helicopter or as complicated as the tactical system for an airborne early warning platform.

The cockpit of an aircraft is a typical location for avionic equipment, including control, monitoring, communications, navigation, weather and anti-collision systems. The majority of aircraft power their avionics using 14- or 28-volt DC electrical systems; however, larger, more sophisticated aircraft (such as airliners or military combat aircraft) have AC systems operating at 400 Hz, 115 volts AC.

Lesson 8 Communication System

Learning Objectives

1. Knowledge objectives:

   A. To grasp words, related terms and abbreviations about communications system on the aircraft.

   B. To grasp the key sentences about communications system.

   C. To know the main components of communications system.

2. Competence objectives:

   A. To be able to read and understand difficult English aviation sentences, capitalized English materials and obtain key aviation information quickly.

   B. To be able to communicate with English speakers about communications system.

   C. To be able to fill in the job cards in English.

3. Quality objectives:

   A. To be able to self-study with the help of aviation dictionaries, the Internet and other resources.

   B. To develop the craftsman spirit of carefulness and responsibility.

Section 1 Aviation Listening & Speaking

1.1 Aviation Listening: listen to the record and fill in the blanks.

1 听力录音　　2 答案及原文

　　As a pilot your main job obviously is to fly the plane. During non critical 1._____phases and 2._____, for example, the 3._____takes over control and you have time to do your regular 30 minutes fuel checks, make sure all 4._____ are working in order and if time permits have a chat with your colleague.

　　As soon as we have completed our approach briefing, we will request and start our initial 5._____towards our destination. As we are descending into lower 6._____and come closer to the airport, the radio communications and the navigation get more complex. There are more aircraft in the vicinity of the airport, so the air 7._____controller is constantly giving instructions and regular frequency changes increase the workload during approach. Terrain is another factor which requires a lot of concentration by pilots especially during bad weather. So therefore 8._____ have come up with a procedure named the sterile cockpit below 10,000 feet, meaning that pilots should fully reduce unnecessary conversation to an absolute minimum and only use standard operating.

Phraseology to ensure a safe cockpit 9._____ in these critical flight phases where a high demand of concentration is required. To avoid 10._____, communications with cabin crew members shall only be established in case of an emergency. Even after landing, the sterile cockpit procedures apply until completion of the parking checklist.

1.2 Aviation Speaking: practice the following dialogue and design your own.

Situation: This is a dialogue between the pilot and the air traffic controller. The pilot reports that the weather is deteriorating and requests radar vector for approach. Practice it and have your own.

Note: PIL=Pilot, CTL= Air Traffic Controller

3 对话录音

Scene 1

CTL: CCA 982, the weather is deteriorating, and RVR runway 31 is less than 1,500 m.

PIL: CCA 982 requests radar vector for approach.

Scene 2

PIL: Beijing Tower CCA 982 ILS established, approaching runway 36R.

CTL: CCA 982, continue approach, and the weather is deteriorating, report outer marker.

PIL: CCA 982 outer marker.

CTL: CCA 982, you are number one, and RVR runway 36 is less than 1,500 m.

PIL: Number one to land CCA 982 . CCA 982 going around.

CTL: Standard procedure, contact 129. 0.

PIL: 129.0 CCA 982.

4 对话译文

Words & Expressions

1. CCA   中国国际航空
2. deteriorate [dɪˈtɪəriəreɪt] v.   变坏；恶化
3. outer marker   远台
4. vector [ˈvektə(r)] n.   航线
5. radar vector   雷达引导
6. Instrument Landing System（ILS）   仪表着陆系统
7. standard procedure   标准复飞程序

Section 2 Aviation Reading

Pre-reading questions:
1. Why is the communications system needed for the aircraft?
2. What makes it possible for the communications for the aircraft?

Communications System

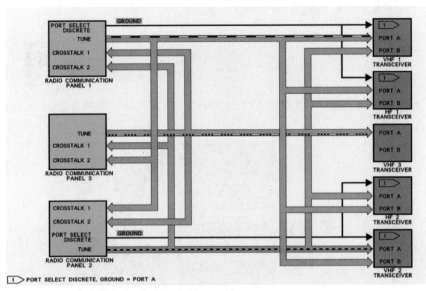

Fig.8-1　Schematic for B737-700 Communications System

Introduction

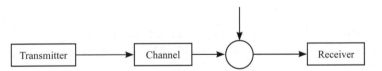

Fig.8-2　Schematics for Communications System

In telecommunication a communications system is a collection of individual communications networks, transmission systems, relay stations, tributary stations, and data terminal equipment (DTE) usually capable of interconnection and interoperation to form an integrated whole. The components of a communications system serve a common purpose, are technically compatible, use common procedures, respond to controls, and operate in union. Telecommunications is a method of communications (for sports broadcasting mass media, journalism, etc.).

The airplane communications systems provide facilities for radio voice communications to ground stations or other aircraft, and distribution of interphone audio, tone signals, announcements, and entertainment audio to personnel and passengers within the airplane.

Two general components are needed for nearly all systems for communications. They are microphones and loudspeakers. Microphones transfer the acoustic information into an electrical signal. Loudspeakers transfer the electrical signal back into acoustical information. Different types of microphone are used in the cockpit. An area microphone, which is usually installed on the ceiling panel, is used by the voice recorder to record the general cockpit sounds; a hand-held microphone is used for announcements to the passengers; and integrated microphones are contained in the oxygen mask or the headset. These are called the boomset.

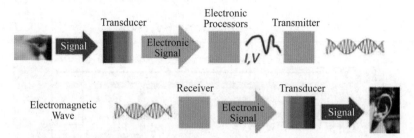

Fig.8-3　An Electronic Communications System Using Electronic Signals

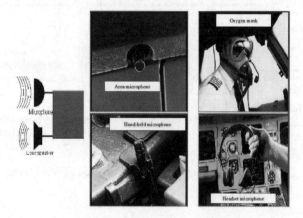

Fig.8-4　Communications Components

Communications Equipment Component Locations

The following are the captain and first officer flight interphone system components:

➤ Boom mic jack.
➤ Headphone jack.
➤ Hand mic jack.
➤ Oxygen mask mic jack.
➤ Control column mic switch.
➤ Flight interphone speaker.
➤ Audio control panel.

The service interphone switch is on P 5 aft overhead panel. The passenger signs panel is on the P 5 forward overhead panel. The controls for the VHF radios are on P 8 aft electronics panel.

The location has service interphone jacks:

➤ P 19 external power panel. (location A)
➤ Electronic rack. (location B)
➤ Right wing refueling slat. (location C)
➤ Right wheel well. (location D)
➤ Left wheel well. (location E)

➢ After entry light panel. (location F)

➢ APU. (location G)

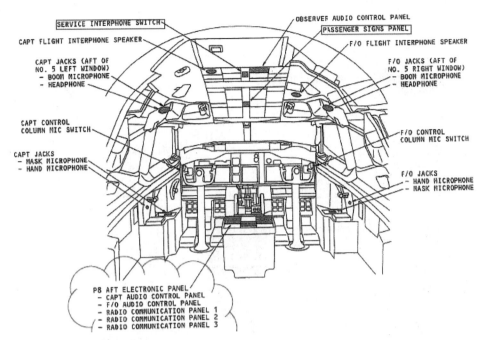

Fig.8-5   Communications Equipment Component Locations-1

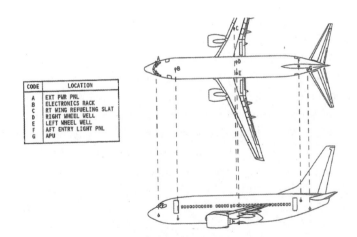

Fig.8-6   Communications Equipment Component Locations-2

Types of Communications Systems

A. Speech Communications Systems

The speech communications systems control the VHF and the HF radios, including systems of High Frequency (HF) communications ad Very High Frequency(VHF) communications. The HF communications system supplies voice communications over long distances. The VHF communications system is used for short distance voice and data communications with ground stations or other aircraft. The Radio Control Panels (RCPs) control and allow for selection of specified frequencies and the mode (AM, SSB) for the HF communications system.

B. HF Communications System

The HF communications system supplies the flight crew with long range voice communications.

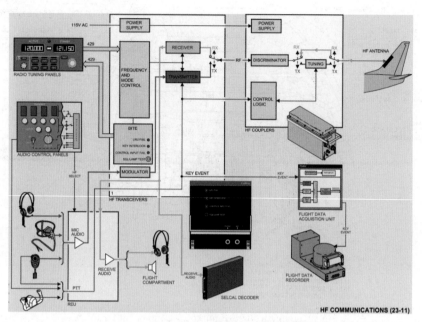

Fig.8-7   Schematic for HF Communications

The HF communications system can be used to communicate between airplanes and between airplanes and ground stations. The HF communications radio uses frequency select and control signals to transmit and receive voice communications. This system operates in the aeronautical frequency range of 2 MHz to 29.999 MHz. Each HF communications system provides Amplitude Modulated (AM) and Single-Sideband (SSB) voice communications between the airplane and ground stations or other airplanes. The HF radio modulates a Radio Frequency (RF) carrier signal with voice audio from the flight interphone system. During the receive mode, the HF radio demodulates the RF carrier signal. This isolates the voice audio from the RF signal. The HF transceiver sends the audio to the flight interphone system.

Fig.8-8   Very High Frequency Communications System

C. Very High Frequency (VHF) Communications System

The VHF frequency range for aircraft communications systems is from 117.975 MHz to 137 MHz for short distance voice and data communications with ground stations or other aircraft. Note that the frequency of 121.5 is an international emergency frequency, which is used, for example, by the emergency locator beacon. The spacing between each communications channel is normally 25 MHz, which means that 760 frequencies, also called channels, are available. In Europe more channels are required in high flight levels, therefore modern systems use a channel spacing of 8.3 MHz which means more than 2000 channels are available.

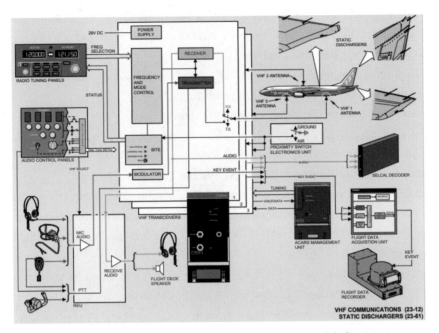

Fig.8-9　Schematic for VHF Communications & Static Dischargers

Use the microphone or headset, RCPs, Audio Control Panels (ACPs) to operate the VHF radio. Next is the receive operation and transmit operation.

Receive operation

➢ Use the RCPs and ACPs to receive transmissions on the VHF radio. On the RCP, push the receiver volume control for the VHF radio.

➢ On the ACPs, push the receiver volume control for the VHF radio. Turn the control to adjust the volume from the VHF radio.

➢ Audio can be heard on the headset and the flight interphone speakers. To hear sound from the flight interphone speakers, push the speaker (SPKR) volume control to turn on the speaker. Turn the control to adjust the volume of sound from the speaker.

➢ As soon as the power is applied to the airplane, the RCPs are on RCP 1 tunes VHF 1 and RCP 2 tunes VHF 2. Push the VHF switch for the VHF radio you want to use. The frequency indicators show VHF radio frequencies (118. 000 to 136.975 MHz).

➢ Listen to the audio from the VHF radio on the speaker or headset. Adjust the volume control switches on the ACPs for a comfortable sound.

➢ Make sure the frequency indicator shows the selected frequency you want to transmit is a valid one.

➢ Push the microphone selector switch on the ACPs for the VHF audio.

Transmit operation

➤ Make sure the active frequency indicator shows the selected frequency you want to transmit is valid.

➤ Push the microphone selector switch on the audio control panel for the VHF radio.

➤ Listen for transmissions on the selected frequency. When the frequency is clear and you want to transmit a message, key the mic and speak into it, then sidetone in the headphone and muted sidetone from the speaker can be heard. When the boom mic or the hand mic are used, the flight interphone system mutes the sidetone to the speaker.

D. Airborne Cabin Telephone System

The airborne cabin telephone system allows passengers and cabin crew to telephone anywhere in the world using International Direct Dialing (IDD).

E. Emergency Locator Transmitter (ELT) System

The ELT system provides an emergency locator signal to aid in search and rescue operations. The ELT transmits on international distress frequencies 121.5, 243.0, and 406.025 MHz.

F. Satellite Communications (SATCOM) System

The SATCOM system provides voice and data communications capability for the flight crew. The SATCOM system can also provide a communications link for passenger voice and data communications. The SATCOM system supplies full-duplex telephone quality voice communications.

The SATCOM system uses 1.5-1.6 GHz frequencies for airplane to satellite links. The SATCOM system supplies voice/data signals over longer distances than the VHF communications system. The SATCOM system provides more reliable communications over oceanic regions than the HF communications system.

Fig.8-10　Emergency Locator Transmitter System

G. Aircraft Communications Addressing and Reporting System (ACARS)

The ACARS provides a ground to air and air to ground airborne communications data link. The system receives digital data, processes it for displaying to the flight crew, and formats digital data of airplane parameters and operations for transmission to ground stations. Messages are coded to identify a specific airplane.

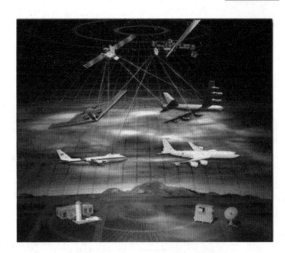

Fig.8-11　SATCOM System

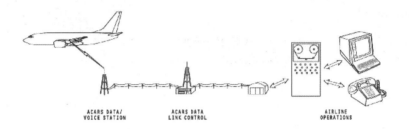

Fig.8-12　ACARS

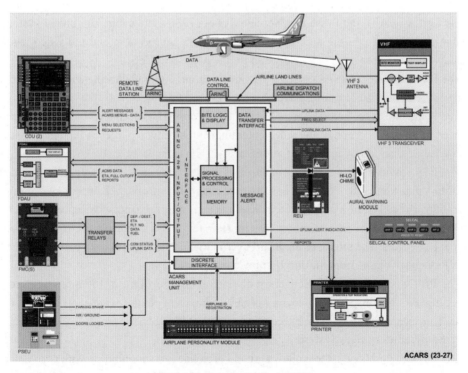

Fig.8-13　Schematic for ACARS

H. Selective Calling (SELCAL) System

The SELCAL system allows a ground operator to alert a specific airplane. Coded tone signals are received with incoming HF or VHF transmissions and when it decoded by the designated airplane, the SELCAL system signals the flight crew with an incoming call.

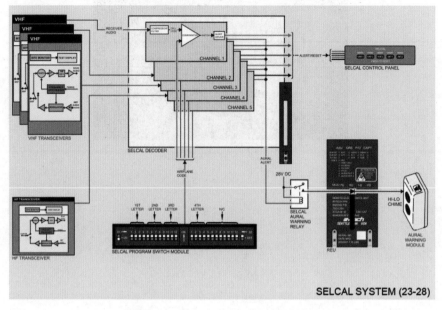

Fig.8-14   Schematic for SELCAL System

I. Advanced Cabin Entertainment/Service System (ACESS)

The ACESS or the Cabin Services System (CSS) includes these systems: the Passenger Address System (PAS), the Passenger Service System (PSS), the cabin interphone system, and the cabin lighting system, the audio passenger entertainment system and the video passenger entertainment system provide audio and video for the ACESS or CSS.

The Passenger Address (PA) system provides a means for announcements to be made to the passengers. Boarding music can be played over the PA system.

The video Passenger Entertainment System (PES) provides selectable video entertainment to passengers. The audio portion of the video entertainment is delivered to each individual passenger seat on selected channels of audio and the PES also may be connected to the PA system speakers.

The PSS provides passengers with control of their reading lights, passenger to attendant call and lavatory functions. The PSS also has the capability of controlling the passenger information signs, muting the call chimes in designated areas and activating the smoke detection alarm.

The audio Passenger Entertainment System (PES) provides selectable pre-recorded music or speech to passengers. The programs are heard over individual headsets at each passenger seat.

J. Passenger Flight Information Display System (PFIDS)

The PFIDS supplies the passengers with data about the flight of the airplane. Such data includes flight routing maps, airline logos, speed, geographical points, external air temperature and altitude. This data is shown through the equipment of the video PES.

K. Service Interphone System

The service interphone system allows ground personnel to communicate with each other from various service stations around the airplane and to the flight crew. They plug a boom/mic headset into jacks located at various points. The flight attendants use the service interphone system to speak with each other and the pilots.

Communications with the flight deck is possible by paralleling the service interphone system with the flight interphone system.

The service interphone system has these components of service interphone jacks, attendant handsets and service interphone switch. The pilots can use the flight interphone system to interface with the service interphone system.

In the flight compartment, select the ACP service interphone function, then push the service interphone mic selector switch and adjust the receiver volume control.

When the handset is removed from the cradle, the attendant handsets automatically connect into the service interphone system. When a headset connects to a service interphone jack, the ground crew can hear the audio. For other stations to hear the ground crew, the service interphone switch must be ON position.

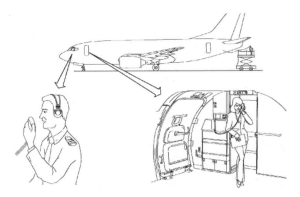

Fig.8-15   Service Interphone System

L. Cabin Interphone System (CIS)

The CIS allows communications from the flight crew to the attendants and attendants to flight crew. Microphones/handsets in the passenger cabin are located in convenient locations for the flight attendants.

M. Ground Crew Call System

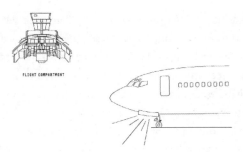

FLIGHT COMPARTMENT

Fig.8-16   Ground Crew Call System-1

The ground crew call system allows personnel in the flight compartment and servicing personnel on the ground outside the airplane to gain each other's attention. They can then establish communications with the service interphone and the flight interphone systems.

The ground crew call system makes call signals between the flight compartment and ground crew. Aural signals remind you of using the service interphone. The ground crew call system has ground call switch on the passenger signs panel, pilot call switch on the external power panel and ground crew call horn.

The system uses the aural warning module for alerts in the flight compartment. Push the GRD CALL switch on the passenger signs panel on P 5 forward overhead panel to call the ground crew. When the crew member uses the switch, a horn in the nose wheel well makes a sound. The ground crew pushes the PILOT CALL switch on P 19 external power panel to call the flight compartment. A call light on the forward overhead panel comes on and there is a chime from the aural warning

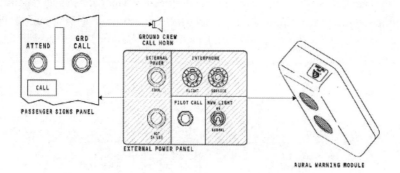

Fig.8-17　Ground Crew Call System-2

N. Cargo Intercom System

The cargo intercom system provides the communications capability required to facilitate cargo handling operations. This system permits communications from the loadmaster stations to the flight compartment and the cargo handler stations.

O. Flight Interphone System

The flight crew uses the flight interphone system to speak with each other and the ground crew. This system interfaces with the communications radios and allows ground to airplane communications to occur. All Push-To-Talk (PTT) microphones, headsets and the oxygen mask microphones are part of the flight interphone system. This system also interfaces with most of the other cabin and crew communications systems. Flight and maintenance crews use the flight interphone system to get access to the communications systems. You can also use it to monitor the navigation receivers. Various navigation systems interface with the flight interphone system, including the VOR/ILS, Marker Beacon, DME and ADF. These systems provide monitoring signals to the flight crew on the headsets or the flight interphone speakers.

The captain and first officer flight interphone system components include the control wheel microphone switch, ACP, hand microphone, oxygen mask, headset and speaker. The observer has the same components without a control column microphone switch, boom mic jack, or speaker. The external power panel has a flight interphone jack for the ground crew.

To talk between the pilot stations, do the following steps on the ACP. First, push down on the FLT INT receive volume switch and turn the switch to adjust the volume. Secondly, push the FLT INT mic selector switch to select the flight interphone system. Thirdly, key a mic switch, and use either the mic switch on the control wheel or the radio-intercom switch on ACP. Finally, talk on a microphone. The audio is heard on the headphones connected at the other stations. The ground crew can listen and talk to the flight crew with a headset connected to the flight interphone jack on

P 19 external power panel.

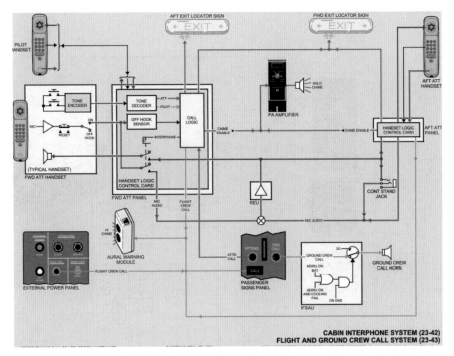

Fig.8-18   Schematic for Cabin Interphone, Flight & Ground Crew Call System

Static Dischargers

Static dischargers eliminate radio receiver interference by discharging static from the fuselage. This is done along the trailing edges of the wings, the horizontal stabilizer and the vertical stabilizer. This limits the interference that can be induced in the radio transceivers.

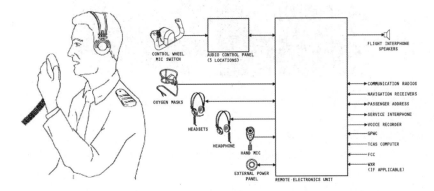

Fig.8-19   Flight Interphone System

Voice Recorder System

The voice recorder system records the flight crew communications and conversation. The voice recorder system keeps the last 120 minutes of audio. The voice recorder unit makes a continuous record of flight crew communications and flight compartment sounds. It makes a continuous record of flight crew communication and flight compartment sounds, and erases the communications data automatically so that the memory stores only recent audio, receiving audio from the Remote Electronics Unit (REU) and the area microphone. The area microphone is in the cockpit voice recorder panel.

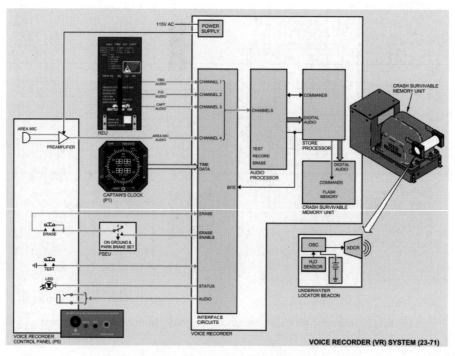

Fig.8-20 Voice Recorder System

Fig.8-21 Voice Recorder & Flight Recorder

Abbreviations & Acronyms

1. Aircraft Communications Addressing and Reporting System (ACARS) 飞机通信寻址与报告系统
2. Audio Control Panel (ACP) 音频控制面板
3. Amplitude Modulated (AM) 调幅
4. Aeronautical Radio Incorporated (ARINC) 航空无线电公司
5. Flight Data Recorder (FDR) 飞行数据记录器
6. Flight Data Recorder System (FDRS) 飞行数据记录系统
7. Light Emitting Diode (LED) 发光二极管
8. Microphone (MIC) 话筒；麦克风
9. Proximity Switch Electronics Unit (PSEU) 接近电门电子组件

10. Radio Communications Panel (RCP)　　　无线电通信面板
11. Remote Electronics Unit (REU)　　　远程电子组件
12. Selective Calling (SELCAL)　　　选择呼叫
13. Single Side Band (SSB)　　　单边带
14. Transmit (XMIT)　　　发射
15. Aircraft Condition Monitoring System (ACMS)　　　飞机状况监控系统
16. Display Electronic Unit (DEU)　　　显示电子组件
17. Digital Flight Data Acquisition Unit (DFDAU)　　　数字飞行数据采集组件
18. Electronic Flight Instrument System (EFIS)　　　电子飞行仪表系统
19. Engine Indication and Crew Alert System (EICAS)　发动机指示与机组警告系统
20. Flight Control Computer (FCC)　　　飞行控制计算机
21. Flight Management Computer (FMC)　　　飞行管理计算机
22. Ground Proximity Warning Computer (GPWC)　　　近地警告计算机
23. Instrument Landing System (ILS)　　　仪表着陆系统
24. Inertial Reference System (IRS)　　　惯性基准系统
25. Navigation Display (ND)　　　导航（数据）显示器
26. Primary Flight Display (PFD)　　　主飞行（数据）显示器
27. International Direct Dialing (IDD)　　　国际直拨电话
28. Emergency Locator Transmitter (ELT)　　　紧急定位发射机
29. Satellite Communications (SATCOM) System　　　卫星通信系统
30. Selective Calling (SELCAL) System　　　选择呼叫系统
31. Advanced Cabin Entertainment/Service System (ACESS)　先进客舱娱乐 / 服务系统
32. Passenger Address System (PAS)　　　旅客广播系统

Words & Expressions

1. telecommunication [ˌtelikəˌmjuːnɪˈkeɪʃ(ə)n] *n.*　　　电信技术
2. communications [kəˌmjuːnɪˈkeɪʃnz] *n.*　　　通信；通讯
3. relay [ˈriːleɪ; rɪˈleɪ] *n.*　　　继电器
4. tributary [ˈtrɪbjətri] *n.*　　　支流
5. terminal [ˈtɜːmɪn(ə)l] *n.&adj.*　　　终端
6. compatible [kəmˈpætəbl] *adj.*　　　可兼容的
7. interphone [ˈɪntəˌfəʊn] *n.*　　　对讲机；内话
8. compatible [kəmˈpætəbl] *adj.*　　　兼容的
9. tone signal　　　单音信号
10. loudspeaker [ˌlaʊdˈspiːkə(r)] *n.*　　　扬声器
11. acoustic [əˈkuːstɪk] *adj.*　　　声音的
12. cockpit [ˈkɒkpɪt] *n.*　　　驾驶舱
13. ceiling panel　　　天花板
14. integrated microphone　　　内置麦克风
15. boom mic　　　吊杆式话筒
16. oxygen mask　　　氧气面罩
17. headset [ˈhedset] *n.*　　　头戴式送受话器
18. boom set　　　吊杆式耳机
19. handset [ˈhændset] *n.*　　　手持式送受话器
20. rack [ræk] *n.*　　　机架

8 单词朗读

9 术语专讲：雷达
（微课）

21. aeronautical [ˌeərəˈnɔːtɪkl] *adj.*	航空的
22. amplitude modulation(AM)	调幅
23. single-sideband (SSB)	单边带
24. channel [ˈtʃæn(ə)l] *n.*	频道
25. locator beacon	定位信标
26. valid [ˈvælɪd] *adj.*	有效的
27. indicator [ˈɪndɪkeɪtə(r)] *n.*	指示器
28. sidetone [ˈsaɪdstəʊn] *n.*	侧音
29. international direct dialing(IDD)	国际通用电话
30. emergency locator transmitter (ELT)	紧急定位发射机
31. satellite link	卫星通信线路
32. capability [ˌkeɪpəˈbɪləti] *n.*	能力，性能
33. full-duplex [ˈfulˈdjuːpleks] *n.*	全双工
34. airborne [ˈeəbɔːn] *adj.*	机载的，空中的
35. parameter [pəˈræmɪtə(r)] *n.*	参数
36. tone signal	单音信号
37. decode [diːˈkəʊd] *v.*	解码
38. mute [mjuːt] *v.*	静音；减弱声音
39. chime [tʃaɪm] *n.*	谐音
40. geographical point	地理点
41. altitude [ˈæltɪtjuːd] *n.*	海拔，高度
42. jack [dʒæk] *n.*	插孔
43. deck [dek] *n.*	甲板，舱面
44. parallel [ˈpærəlel] *v.*	并联
45. volume [ˈvɒljuːm] *n.*	体积，音量
46. cradle [ˈkreɪd(ə)l] *n.*	支架，插簧
47. aural [ˈɔːrəl] *adj.*	听觉的
48. flight attendant	空中服务员
49. marker beacon	指点信标
50. static discharger	静电放电刷
51. interference [ˌɪntəˈfɪərəns] *n.*	干扰
52. radio transceiver	无线电收发机

Exercises

2.1 Choose the best answer from A, B, C, D options.

10 答案

1. electronic
   A. 通电   B. 电路   C. 电子   D. 中子

2. radio
   A. 雷达   B. 无线电   C. 录像机   D. 音频

3. telecommunication
   A. 联通   B. 通信   C. 电报   D. 电缆

4. helicopter
   A. 喷气式飞机   B. 无人机   C. 战斗机   D. 直升机

5. Alternating Current(AC)

    A. 直流电   B. 交流电   C. 音频控制面板   D. 无线电通信面板

6. 插孔

    A. jack   B. jet   C. rack   D. switch

7. 甚高频

    A. Automatic Direction Finder(ADF)   B. Radio Frequency(RF)

    C. High Frequency(HF)   D. Very High Frequency(VHF)

8. 头戴式送受话器

    A. handset   B. headset   C. handphone   D. headphone

9. 收发机

    A. transformer   B. receiver   C. transceiver   D. transmitter

10. 客舱内话系统

    A. service interphone system   B. flight interphone system

    C. cabin interphone system   D. cargo intercom system

2.2 Find the best option to paraphrase the sentence.

1. Two general components are needed for nearly all systems for communications.

    A. Two general components are needed for some systems for communications.

    B. Two general components are needed for almost all systems for communications.

    C. Two general components are unnecessary for nearly all systems for communications.

    D. Two general components are needed for none of systems for communications.

2. The HF communications system supplies the flight crew with long range voice communications.

    A. The HF communications system provides the flight crew with long range video communications.

    B. The HF communications system provides the flight crew with long range voice communications.

    C. The VHF communications system supplies the flight crew with long range voice communications.

    D. The HF communications system supplies short range voice communications with the flight crew.

3. The SELCAL system allows a ground operator to alert a specific airplane.

    A. The SELCAL system prohibits a ground operator to alert a specific airplane.

    B. The SELCAL system allows a ground operator to alert a special airplane.

    C. The SELCAL system permits a ground operator to alert a specific airplane.

    D. The SELCAL system allows a ground operator to alert all airplanes.

4. The sequence of topics in the Handbook is intended to increase in-flight usefulness.

    A. The order of topics in the Handbook is intended to increase in-flight usefulness.

    B. The sequence of topics in the Handbook is intended to increase in-flight useless.

    C. The sequence of topics in the Handbook is pretended to increase in-flight usefulness.

    D. The sequence of topics in the Handbook is intended to decrease in-flight usefulness.

5. It erases the communications data automatically so that the memory stores only recent audio.

    A. It erases the communications data automatically so that the memory remembers only recent video.

B. It records the communications data automatically so that the memory stores only recent audio.

C. It erases the communications data automatically so that the memory stores only recent video.

D. The communications data is erased automatically so that the memory stores only recent audio.

2.3 Change the following components specified in lower case and translate them into Chinese.

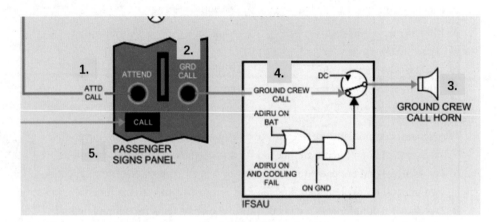

1.	
2.	
3.	
4.	
5.	

Section 3 Aviation Translation: translate the following sentences into Chinese

1. CAUTION: REMOVE THE AERODYNAMIC FILLET SEAL CAREFULLY WITH THE SEALANT REMOVAL TOOL. IF NOT, DAMAGES TO THE AIRPLANE SKIN OR THE COAXIAL CABLE CAN OCCUR.

11 航空安全警示语
（文本）

2. WARNING: MAKE SURE THAT PERSONNEL STAY A MINIMUM OF 10 FT (3 M) AWAY FROM THE VERTICAL STABILIZER WHEN THE HF COMMUNICATIONS SYSTEM TRANSMITS. RF ENERGY FROM THE HF ANTENNA CAN CAUSE INJURIES TO PERSONNEL.

Section 4 Aviation Writing

Liu Shuai is an aircraft mechanic. He is inspecting the plane and finds some faults.

Please help him write down fault descriptions. Some related words, phrases & terms and key sentences are offered as follows.

Key words, phrases & terms:

1.	高频通信	High Frequency (HF) Communications
2.	高频收发机	HF transceiver
3.	高频天线耦合器	HF antenna coupler
4.	键控互锁	key interlock
5.	调谐中	tune in progress
6.	临近电门电子组件	Proximity Switch Electronics Unit (PSEU)
7.	头戴式耳机	headphone
8.	按压通话	push-to-talk (PTT)
9.	甚高频通信	Very High Frequency (VHF) Communications
10.	甚高频天线	VHF antenna
11.	遥控电子组件	Remote Electronics Unit (REU)
12.	无线电通信面板	Radio Communication Panel (RCP)
13.	选呼控制面板	SELCAL control panel
14.	选呼程序开关组件	SELCAL program switch module
15.	选呼音响警告继电器	SELCAL aural warning relay
16.	旅客广播	Passenger Address (PA)
17.	旅客广播放大器	PA amplifier
18.	乘务员话筒	attendant headset
19.	手持话筒	hand microphone
20.	发光二极管显示器	LED display

Key sentences:

1. 旅客广播系统在驾驶舱进行广播时变音 / 断续 / 音量有问题。
   The PA system distorted/intermitted/volume problem from the flight compartment.
2. 后登机门灯面板插孔。
   Jack at aft entry light panel.
3. 电子设备架插孔。
   Jack at EE rack.
4. 左轮舱插孔。
   Jack at left wheel well.
5. 右翼加油缝翼插孔。
   Jack at right wing refueling slat.
6. 电子设备舱插孔。
   Jack in the electronic equipment bay.
7. 乘务员 / 地面人员呼叫驾驶舱灯不亮。
   Attendant/ ground crew to flight compartment call light does not come on.
8. 驾驶舱呼叫音不响。
   Chime does not sound in the flight compartment.
9. 呼叫喇叭。
   Call horn.
10. 不响。
    Does not sound.

Task: Liu Shuai has three writing tasks. The first one is that there is no noise for the earphone, the second is to jack at external power receptacle panel, and the last one is to jack at right wheel well. Please finish the writing tasks for him.

Warning Case

Tenerife Disaster

12 警示案例——
特内里费空难

On March 27th, 1977, two Boeing 747s were waiting to take off on the runway and could not see each other because of heavy fog at Las Palma airport. The poor communications between the pilots and the air traffic controller led to the two planes' taking off at the same time and a tragic collision on the runway. A total of 583 people were killed in the incident. Among them, 257 people on KLM (K.L.M. Royal Dutch Airlines) were all killed, 61 people on Panam (Panamerican World Airways) survived miraculously.

Warning Tips: there are many factors causing this incident, one of which is that the controller did not hear "we are taking off" with Dutch accent clearly, but "we are at take off", so he replied: "OK, stand by for taking off, wait for us to inform you!" The KLM crew only heard the word "OK".

Lesson 9 Electrical Power System

Learning Objectives

1. Knowledge objectives:

   A. To grasp the words, related terms and abbreviations about the electrical power system.

   B. To grasp the key sentences about the electrical power system.

   C. To know the function and main components of the electrical power system.

2. Competence objectives:

   A. To be able to read and understand frequently-used & complex sentence patterns, capitalized English materials and obtain key information quickly.

   B. To be able to communicate with English speakers freely.

   C. To be able to fill in the job cards in English.

3. Quality objectives:

   A. To be able to self-study with the help of aviation dictionaries, the Internet and other resources.

   B. To develop the craftsman spirit of carefulness and responsibility.

Section 1 Aviation Listening & Speaking

1.1 Aviation Listening: listen to the record and fill in the blanks.

1 听力录音   2 答案及原文

As the 1._____passes through clouds and due to the friction of the air, the plane hull gets charged with electrons. The 2._____electricity has the tendency to accumulate at sharp edges and peaky objects. Similar as if you would walk across a velour carpet, due to the friction of your shoes with the carpet the electrons get picked up and your body is statically charged with 3._____. And we've all had the 4._____ as you touch a 5._____object or someone with a different static load and you get that little electric 6._____.

The problem is that radio and 7._____antennas on an airplane are fairly edgy and peaky meaning that the electrons would accumulate there and create a nasty static noise making 8._____ barely understandable, and the navigation can also be jeopardized by the magnetic field created by the static electricity.

Therefore airplane manufacturers came up with an easy solution by installing little static wicks at the most prominent sharp edges like 9._____, flap track fairings, rudders and elevators, so that the electrons have an easier way of 10. _____from the plane without disturbing the radio and navigation antennas.

1.2 Aviation Speaking: practice the following dialogue and design your own.

3 对话录音

4 对话译文

Situation:

The ground mechanic and the pilot on board are having a dialogue when the parking brake valve deactivation occurred in the closed position. Practice it and have your own.

Note: PIL=Pilot, GND=Ground Mechanic

Scene 1

GND: Captain, I have installed the downclock pins on the landing gear.

PIL:　Roger, copy that.

GND: The plane has been supplied electrical power, and the wheels have had chocks installed around them. Please release the parking brake.

PIL:　Roger. Brake is released. It seems that the parking breake valve doesn't work.

GND: Roger. We will check the MEL for dispatching the aircraft.

Scene 2

GND: Cockpit, I have checked the relevant MEL item. I will follow the MEL maintenance procedure to open these circuit breakers and install safety tags.

PIL:　Roger. Go ahead.

GND: I have disconnected the electrical connector from the parking brake valve in the main wheel bay.

PIL:　Roger. Confirm that.

GND: I will move the position override lever on the parking brake valve to the PQS 2 (closed position) manually.

PIL:　Roger.

GND: I've removed all the safety tags and closed these circuit breakers. The antiskid INOP lights are on. I'll set the autobrake switch to the "OFF" position. Please confirm that.

PIL:　Roger. Confirm the autobrake switch on the "OFF" position.

GND: I will inform the SOC. The parking brake valve is in deactivation.

PIL:　Roger. We will carry on the NON-ANTISKID procedure.

Words & Expressions

1. chock [tʃɒk] *n.*　　　　　　　　　轮挡

2. Minimum Equipment List (MEL)　　　最低设备清单
3. safety tag　　　安全标签
4. the position override lever　　　位置超越操纵杆
5. Systems Operation Center (SOC)　　　系统运行中心
6. deactivation [dɪˌæktɪ'veɪʃən] *n.*　　　无效，不活动

Section 2 Aviation Reading

Pre-reading questions:

1. Where does electrical power come from?
2. Why does electrical power is needed on the aircraft?

 5 课文朗读　6 课文译文

7 飞机上的电源系统（微课）

Electrical Power System

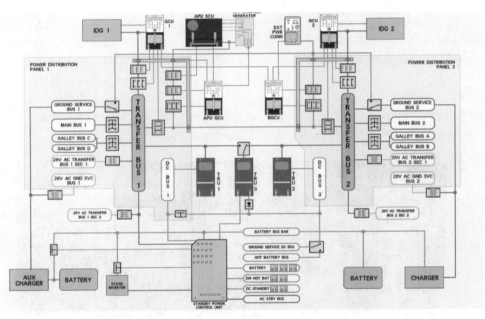

Fig.9-1　Schematic for B737-700 Electrical Power System

The electrical power system is used for the control, operation, and indication of the various airplane systems on the ground and in-flight. It is designed to supply airplane systems with AC and DC power. The electrical power is obtained from the battery, generator and ground support equipment. It is controlled and monitored prior to distribution as 115 volt AC, 28 volt AC and 28 volt DC supply to using systems.

Main Components & Their Locations

Main Components
Constant Speed Drive (CSD)

The CSD is located on the left side of the engine and the front of the accessory gearbox. The purpose of CSD is to convert the variable speed of the engine to a constant speed for the driven generator to produce AC power at constant frequency.

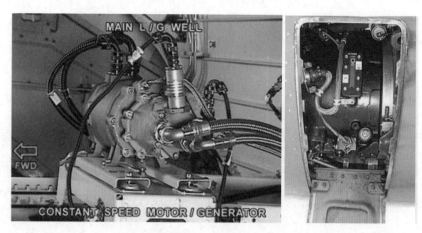

Fig.9-2　Constant Speed Drive (CSD)

Integrated Drive Generators (IDGs)

The IDGs are on the forward face of the engine accessory gearbox. The air/oil cooler is on the engine fan case. Each engine has an AC generator. The constant-speed drive unit (or other system) is the link between the generator and the engine. The drive unit has its own oil system for cooling and lubrication and should be checked in the sight glass before flight. This oil system is independent of engine oil system.

IDGs will auto-disconnect with a high oil temperature, thus no IDG oil temperature gauges on the gen drive panel. They are cooled by both fan air and a fuel-oil heat exchanger.

Fig.9-3　Integrated Drive Generator(IDG, NG'S only)

Battery is a 36 ampere-hour, 24 volt, 20 cell, Nickel-cadmium battery and should provide 30 minutes (20 mins 1/200's) of standby power if all other generators fail. The battery is located in the electronic compartment, left side just forward of the E2 rack. The purpose of the battery is to provide DC power to critical airplane systems in absence of normal DC supply from the transformer rectifier. It is also used as a backup power for the AC system control and protection and for starting the APU.

Fig.9-4 Battery

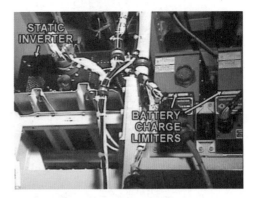

Fig.9-5 Static Inverter

APU battery is primarily used for starting the APU but also works in parallel with the main battery to provide 45 mins of standby power. One of its best applications is that power is retained on the Captains EFIS with the loss of all generators, similar to latest build classics.

Aux battery is a reserve battery on the NG which is normally isolated unless the main battery is powering the standby system when it operates in parallel with the main battery. The aux battery combined with the main battery will provide 60 minutes of standby power.

Transformer Rectifier Units (TRUs)

Three TRUs are located in the electronic compartment on E3 rack. The purpose of the 3 TRUs is to convert 115 volt AC, 400Hz, 3-phase power to 28 volt DC power for use by the airplane's systems.

Fig.9-6 TRU

The APU starter-generator is on the APU gearbox.

Locations of the Main Components

Electrical power components are located in the following areas: flight compartment, external power panel, forward lower fuselage, EE compartment, engine and APU compartment.

Battery, battery charger, Generator Control Units (GCUs), Bus Power Control Unit (BPCU), Power Distribution Panels (PDPs), Start Converter Unit (SCU) are in the EE compartment. The GCUs and BPCU supply BITE for the AC power and external power systems. The battery is located in the electronic equipment compartment, left side just forward of the E2 rack. The static inverter, battery charger, transformers, rectifiers, and a lot of relays are located on electronic equipment E3 rack.

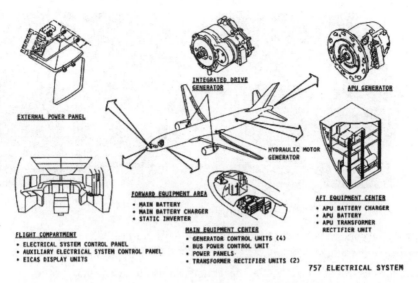

Fig.9-7　Electrical Power Components Locations

Operating Principles of Electrical Power

A standby AC and DC system gives normal and emergency power. The AC power is supplied by two engine driven generators for normal in-flight operation. A generator, driven by the auxiliary power unit, can supply all power for ground or flight operation. Power can be supplied on the ground.

Fig.9-8　Gen Bus Panel-Classic & NG

The AC power is supplied by two engine driven generators for normal in-flight operation. They produce AC power required by the airplane systems primarily for flight operation. The AC generator is rated at 50 KVA (kilovolt-amperes), 120/208 volts, 400Hz. The unit is without slip-rings, commutator, or brushes on either the main generator or the exciter. A complete generator assembly consists of an exciter generator, a rotating rectifier and a main generator.

The TRUs convert 115V AC into 28V DC to supply their associated DC 1 and DC 2. Bus 1 then supplies the DC BAT bus. The bus can charge the batteries or receive power from the batteries as a backup supply, if no other power sources are available. The electrical system also includes two Essential (ESS) Buses. One is the AC ESS bus fed by AC bus 1 and another is the DC ESS bus fed by DC bus 1. These buses are used to supply the most critical A/C systems.

A generator, driven by the auxiliary power unit, can supply all power for ground or flight operation. The entire electrical network can also be supplied by it. On the ground, power can be supplied through the external power (AC) receptacle. The aircraft electrical network can be supplied by an external power source. The DC power is supplied from the battery or conversion from the AC power. The AC and DC power is distributed to the various systems through the left and right load control centers in the flight compartment. In emergency configuration with emergency generator not available, BAT 1 supplies the AC ESS bus via the static inverter and BAT 2 supplies the DC ESS bus.

Any one of the power sources can supply the entire electrical network. The control and indication of the electrical system is from P 5 overhead panel in the flight compartment.

Different Types of Electrical Power

AC Power

The electrical power system generates and controls 115/200-volt, 3-phase, 400-Hz AC power. Two IDGs supply main AC power with the capacity of 90KVA. An APU starter-generator provides in-flight backup to the IDGs (with the capacity of 90KVA below 32,000feet/9,753 meters, and goes down to 66 KVA at 41,000 feet/12,496 meters). The external power supply AC power with the capacity of 90KVA. The IDGs and APU starter-generator supply a 3-phase, 115/200 volts (nominal) at 400Hz. The static inverter is powered from the hot battery bus with a single phase, 115V AC output to the AC standby bus.

DC Power

The DC power is supplied from the battery or conversion from the AC power. The airplane also has these DC power sources by battery and battery charger. The main battery and its charger provide a backup source for the standby power system if other sources do not operate. Three TRUs change 115V AC to 28V DC.

Standby Power

If the normal power is lost, the standby power system supplies a minimum of 30 minutes of AC and DC power necessary to maintain safe flight. The battery supplies DC power. The static inverter uses battery power to make AC power.

Subsystems of Electrical Power

The system has automatic and manual controls and protection. Built-in-test Equipment (BITE) and alternate source selection make the system reliable and easy to maintain. The electrical power has the following subsystems:

- ➢ Generator drive.
- ➢ AC generator.
- ➢ DC generator.
- ➢ External power.
- ➢ AC electrical load distribution.

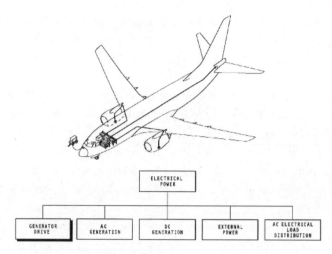

Fig.9-9   Subsystems of Electrical Power

Control & Indications

Modules on P 5 overhead panel supply manual control, indication and DC and standby power system Built-in-test Equipment (BITE). P 6 and P 18 have many circuit breakers and relays. Flight crew cycling (pulling and resetting) of a circuit breaker to clear a non-normal condition is   not recommended, unless directed by a non-normal checklist. The electrical power for the ground operation of all electrical loads is supplied by an external power system supplied through P 19 external power receptacle on the right side of the airplane forward of the nose wheel well.

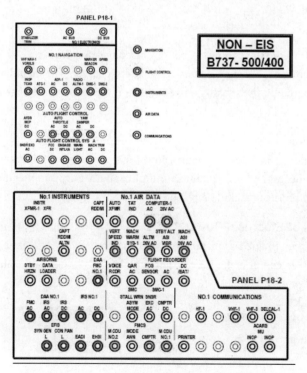

Fig.9-10   B737-3/4/500 Circuit Breaker

Fig.9-11　P 6　　　　Fig.9-12　Circuit Breakers behind P 6　　　　Fig.9-13　P 18

The control and indication of the electrical system is on P 5 overhead panel in the flight compartment. Electrical meters, battery and galley power module (P 5-13), generator drive and standby power module (P 5-5), and AC system, generator and APU module (P 5-4) let you control the electrical system manually.

Fig.9-14　Gen Drive & Standby Power Module - NG

Fig.9-15　P 5-13 & P 5-4

Electrical Meters, Battery and Galley Power Module (P 5-13)

The P 5-13 has the following functions. First, see the parameters for AC and DC components or buses. Then, connect battery power to electrical buses with the battery switch. Thirdly, supply and remove power to the galleys with the galley switch.

It also has the BITE functions of stowing DC and standby power system failure indications, monitoring the DC and standby power, saving fault messages in memory and showing fault messages on the LED alphanumeric display.

The AC meter selector is rotary selector with seven positions, including GEN1&2, INV, APU GRD, PWR, STBY PWR and TEST. Each position stands for an AC power source or AC bus except the TEST position. The alphanumeric display gives the information of voltage, load, and output frequency for the three generators. When the STBY PWR position, GRD PWR position, INV position are selected, it only shows voltage and frequency.

The DC meter selector is also a seven positions rotary selector. Each position stands for a DC power source or DC bus. When the selector is put in TR 1, TR 2, TR 3 position or the BAT position, voltage and load are shown. While the selector is in the BAT BUS or STBY PWR positions, only voltage is shown. The AC and DC selectors must be in TEST position to use the P 5-13 BITE.

(a)AC & DC Metering panel – Classic　　　　　(b) AC & DC Metering panel – NG

Fig.9-16　　　AC & DC Metering Panel

Generator Drive and Standby Power Module (P 5-5)

The P 5-5 has DRIVE light (IDG low oil pressure indication), STANDBY PWR OFF light(indication that either the battery bus or standby bus do not have power), DISCONNECT switch (generator drive disconnect switch), and STANDBY PWR switch.

When the IDG oil pressure is less than the operation limit, the amber DRIVE light is on. When the respective engine start lever is in the idle position, the DISCONNECT switch operates the disconnect mechanism for its IDG. This removes the engine Accessory Gearbox(AGB) power from the IDG. There is one switch for each IDG. Each switch is spring-loaded to the NORMAL position. The DISCONNECT position is a momentary position. The cover is a guard for the switch in the NORMAL position. Break-away wire usually keeps the cover down.

The amber STANDBY PWR OFF Light comes on, when any of the AC standby bus, DC standby bus and Battery bus do not have power. This switch gives manual control of the AC and DC standby power bus sources. It is a three-position switch, usually in the AUTO position. The cover is guarded in the AUTO position.

AC System, Generator and APU Module (P 5-4)

The lights of GRD POWER AVIAILABLE, TRANSFER BUS OFF, SOURCE BUS OFF, GEN OFF BUS, and APU GEN OFF BUS is located in the upper part of the AC system, generator and APU module (P 5-4). The manual controls of ground power switch, engine generator switch, APU generator switch, and bus transfer switch are also in this area.

If the external AC power is connected and the quality is good, the bright blue GRD POWER AVAILABLE light is on. Use the ground power switch to control external power to the AC transfer buses.

If the AC transfer bus does not have power, the amber TRANSFER BUS OFF light is on. Each transfer bus has one light. If an AC transfer bus is not energized by the selected source, the amber SOURCE BUS OFF light comes on. The left SOURCE BUS OFF light is related to IDG 1 (GEN 1 switch), APU (left APU GEN switch) and external power (ground power switch). The right SOURCE BUS OFF light is related to IDG 2 (GEN 2 switch), APU (right APU GEN switch), and external power (ground power switch).

The SOURCE BUS OFF light does not show the AC transfer bus is de-energized. When Generator Control Breaker (GCB) 1 trips, the left SOURCE BUS OFF light comes on in fight. However, the bus transfer function lets IDG 2 power AC transfer bus 1.

When the GCB is open, the blue GEN OFF BUS light comes on, and when the GCB closes, the light goes off. This shows IDG is not a power source in use. The blue APU GEN OFF BUS light is on to show that APU is running, but its generator is not a power source in use. When the APU is running and the Auxiliary Power Breaker (APB) is open, the light is on. When the APB closes or you shut down the APU, the light goes off.

Manually control IDG power source selection with the engine generator switches. Each switch has ON, OFF and NEUTRAL positions, which is always spring-loaded to NEUTRAL position. Put a generator switch momentarily to the ON position to make IDG power its AC transfer bus. If IDG power quality is good, the electrical power system first removes the present power source, then the generator breaker closes and the IDG supplies power.

The APU engine generator switches operate like the engine generator. There are two Bus Tie Breakers (BTBs) so there are two switches.

Manual control the BTBs and the DC bus tie relay with the bus transfer switch. The switch has two positions, and it is usually in the AUTO position. A cover is guarded in the AUTO position. The switch in the AUTO position, the BTBs and the DC bus tie relay work automatically as necessary,bus in the OFF position, the BTBs and the DC bus tie relay open.

Protection of the Electrical Power System

The electrical power system has automatic control to protect the system from source failure or load failure. The left and right Generator Control Unit (GCU1& GCU2), APU Generator Control Unit (AGCU), Bus Power Control Unit (BPCU), Standby Power Control Unit (SPCU) and Start Converter Unit (SCU) are the LRUs which can do this job.

The GCU monitors the system to control and protect the IDGs. APU and the SCU work together to control and protect the APU starter generator. The BPCU controls and monitors the use of external power when the quality of the external power is out of limits.

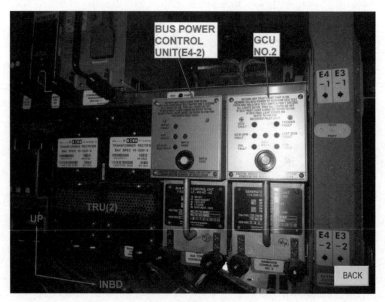

Fig.9-17   BPCU&GCU

Electrical Power Distribution System

The purpose of electrical power distribution system is to provide and control generated AC and DC power for use by the various airplane systems. It consists of 115 volt AC, 28 volt AC and 28 volt DC power obtained from generators and batteries.

The transfer bus 1&2, and ground service bus 1&2 receive power directly from AC power sources. System logic automatically removes loads (load shed) to prevent the overload of the AC power source.

The DC bus 1&2 and the battery bus receive power directly from the TRUs. The hot battery bus and the switched hot battery bus receive power directly from the battery or the battery charger.

The external power, APU starter-generator, and IDGs supply power to the AC transfer buses. The system design makes sure that two AC power sources can not supply power to the same transfer bus at the same time. However, one AC power source can supply power to both transfer buses through the Bus Tie Breakers (BTBs).

Each transfer bus supplies power to Galleys (as many as two), Main Bus, Ground Service Bus, and Transformer Rectifier Unit (as many as two).

Each ground service bus receives power in two ways. First, the AC transfer bus on that side has power. Second, the ground service switch on the forward attendant's panel is in the ON position and external power is connected to the airplane. The two ground service transfer relays control the selection of the power source. The main buses and the galley buses receive power from their respective AC transfer bus. Load shed relays remove the power to the buses when the loads exceed operating limits, which protects the AC power source from overload.

Operation of Electrical Power

The AC and DC power is distributed to the various systems through the left and right load control centers in the flight compartment. Use switches on the forward P 5 overhead panel or the forward flight attendant's panel to operate the electrical system.

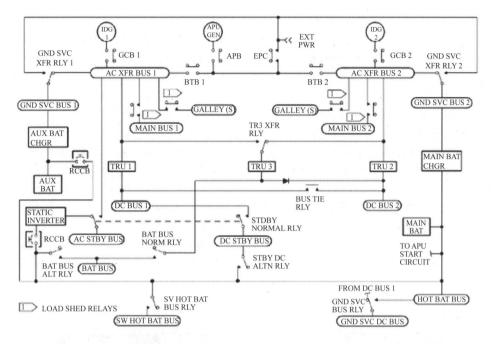

Fig.9-18   Electrical Power Distribution System

Operation of Electrical Meters, Battery and Galley Power Module (P 5-13)

Put the BAT switch to the ON position to energize the switched hot battery bus, battery bus, static inverter, AC standby bus, DC standby bus, P 5-13 alphanumeric display with battery power.

Use the GALLEY switch to control power to all galleys.

Use the DC and AC selectors, and the alphanumeric display to monitor the electrical power system power sources.

Operation of Generator Drive and Standby Power Module (P 5-5)

The generator drive disconnect switch operates the disconnect mechanism for its IDG. This removes engine accessory gearbox power from the IDG. The engine start lever must be in the idle position for the disconnect function to operate.

The standby power switch manually controls the AC and DC standby power bus sources. In the AUTO position, the AC standby bus receives power from DC bus 1 and the DC standby bus from DC bus 1 if these sources are available. If the sources are not available, the AC standby bus receives power from the static inverter and the DC bus receives power from the battery.

The standby power switch has the following functions in the other two positions. First, de-energize the AC standby bus and the DC standby bus (OFF position); secondly, energize the AC standby bus with battery power through the static inverter; thirdly, energize the DC standby bus with battery power (BAT position).

Operation of AC System, Generator and APU Module (P 5-4)

Use the ground power switch to control external power to the AC transfer buses. If the ground source is connected and quality is good, the blue GRD POWER AVAILABLE light comes on. When the ground power switch is put to the ON position, both AC transfer buses receive power. Any initial power sources are removed before the transfer buses receive external power.

The bus transfer switch manually control the BTBs and the DC bus tie relay. In the AUTO position, the Bus Tie Breakers(BTBs) and the DC bus tie relay operate automatically as necessary. In the OFF position, the BTBs open and isolate the AC transfer buses from each other. The DC bus tie relays will also open, and the BTB trip circuits are reset.

Use the GEN 1 and GEN 2 switches to supply IDG power to an AC transfer bus. When the switch is put temporarily to the ON position, any initial power sources will be removed. The AC transfer bus on that side goes to IDG power.

Use the APU GEN switches to supply power to the AC transfer buses. There are two switches, so there are two BTBs. If both AC transfer buses do not have power initially, both AC transfer buses will receive APU power with operation of just one APU GEN switch. If the AC transfer buses do have power initially, then only the AC transfer bus on the same side as the APU GEN switch will energize with APU power.

The blue APU GEN OFF BUS light comes on when the APU is ready to supply electrical power. Use the ground service switch to supply external power to ground service bus 1 and 2 with external power connected. This makes it possible to supply electrical power for cabin servicing without going into the flight compartment.

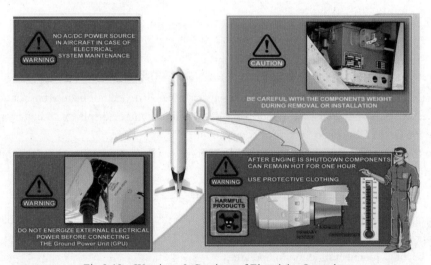

Fig.9-19　Warnings & Cautions of Electricity Operation

Abbreviations & Acronyms

1. Power (PWR)　　　　　　　　　　电力
2. Alternative Current（AC）　　　　交流电
3. Direct Current（DC）　　　　　　直流电
4. Generator (GEN)　　　　　　　　发电机
5. Ground (GND)　　　　　　　　　接地
6. Constant Speed Drive (CSD)　　　恒速驱动
7. Accessory Gearbox (AGB)　　　　附件齿轮箱
8. Integrated Drive Generator (IDG)　整体驱动发电机

9. Disconnect (DISC)　　　　　　　　　　断开
10. Standby (STDBY)　　　　　　　　　　待机
11. Transformer (XFR)　　　　　　　　　　变压器
12. Transformer Rectifier Unit (TRU)　　　变压器整流组件
13. Charger (CHGR)　　　　　　　　　　　充电器
14. Generator Control Unit (GCU)　　　　　发电机控制组件
15. Bus Power Control Unit (BPCU)　　　　总线电源控制组件
16. Power Distribution Panel (PDP)　　　　配电板
17. Start Converter Unit (SCU)　　　　　　起动转换组件
18. Built in Test Equipment (BITE)　　　　机载自测设备
19. Relay (RLY)　　　　　　　　　　　　继电器
20. Alternate (ALTN)　　　　　　　　　　交替的
21. Generator Control Breaker (GCB)　　　发电机控制跳开关
22. APU Breaker (APB)　　　　　　　　　APU 跳开关
23. Bus Tie Breaker (BTB)　　　　　　　　汇流条连接跳开关
24. Generator Control Unit (GCU)　　　　　发电机控制组件
25. APU Generator Control Unit(AGCU)　　APU 发电机控制装置
26. Standby Power Control Unit（SPCU）　 备用电源控制组件
27. APU Starter-Generator (ASG）　　　　 APU 起动发电机

Words & Expressions

1. electrical power　　　　　　　　　　　电源
2. battery ['bætri] *n.*　　　　　　　　　电瓶
3. generator ['dʒenəreɪtə(r)] *n.*　　　　发电机
4. distribution [ˌdɪstrɪ'bjuːʃn] *n.*　　　分配
5. volt [vəʊlt] *n.*　　　　　　　　　　伏特
6. alternating current (AC)　　　　　　　交流电
7. direct current (DC)　　　　　　　　　直流电
8. converter [kən'vɜːtə(r)] *n.*　　　　　整流器
9. integrate ['ɪntɪgreɪt] *adj.*　　　　　集成的
10. ampere ['æmpeə(r)] *n.*　　　　　　　安培
11. cell [sel] *n.*　　　　　　　　　　　电池
12. nickel ['nɪk(ə)l] *n.*　　　　　　　　镍
13. cadmium ['kædmiəm] *n.*　　　　　　镉
14. accessory [ək'sesəri] *n.*　　　　　　附件
15. frequency ['friːkwənsi] *n.*　　　　　频率
16. standby ['stændbaɪ] *adj.*　　　　　　备用的
17. compartment [kəm'pɑːtmənt] *n.&v.*　舱，隔舱，分隔
18. backup power　　　　　　　　　　　备用电源
19. starter-generator　　　　　　　　　　起动发电机
20. meter ['miːtə(r)] *n.*　　　　　　　　电表
21. transformer rectifier　　　　　　　　　变压整流器
22. hertz [hɜːts] *n.*　　　　　　　　　　赫兹
23. phase [feɪz] *n.*　　　　　　　　　　相
24. static inverter　　　　　　　　　　　静变流机

8 单词朗读

9 术语专讲：继电器
（微课）

25. overhead panel（OHP）   头顶板
26. receptacle [rɪ'septəkl] *n.*   插座
27. charger ['tʃɑːdʒə(r)] *n.*   充电器
28. relay ['riːleɪ] *n.*   继电器
29. alternate [ɔːl'tɜːnət] *adj.*   备用的
30. rate [reɪt] *v.*   定额，定……的等级
31. essential [ɪ'senʃl] *a.*   本质的，主要的
32. bus [bʌs] *n.*   汇流条
33. integrated drive generator (IDG)   整体驱动发动机
34. circuit breaker   电路跳开关
35. nose [nəʊz] *adj.*   前部的
36. module ['mɒdjuːl] *n.*   组件
37. built-in-test   内装测试，自检
38. remote control   遥控，远程控制
39. alphanumeric [ˌælfənjuː'merɪk] *adj.*   字母数字的
40. voltage ['vəʊltɪdʒ] *n.*   电压（量），电位差
41. output ['aʊtpʊt] *n.&v.*   输出
42. energize ['enədʒaɪz] *v.*   通电；供电
43. momentary ['məʊməntri] *adj.*   瞬时的
44. failure ['feɪljə(r)] *n.*   故障
45. load shed relay   卸载继电器

Exercises

2.1 Choose the best answer from A, B, C, D options.

1. relay
   A. 继电器   B. 充电器   C. 整流器   D. 变流器

   10 答案

2. phase
   A. 词组   B. 面貌   C. 节奏   D. 相
3. module
   A. 模型   B. 组件   C. 调整   D. 情态动词
4. receptacle
   A. 接受   B. 接待   C. 插座   D. 收据
5. energize
   A. 通电   B. 能量   C. 精力旺盛的   D. 无力
6. 发电机
   A. motor   B. generator   C. generation   D. generate
7. 汇流条
   A. bush   B. brush   C. bar   D. bus
8. 电压
   A. current   B. volt   C. voltage   D. electric resistance
9. 备用电源
   A. backup power   B. back power   C. uphold power   D. setup power
10. 变压器
    A. transformer   B. transmitter   C. transceiver   D. transaction

2.2 Find the best option to paraphrase the sentence.

1. When the IDG oil pressure is less than the operation limit, the amber DRIVE light is on.

    A. When the IDG oil pressure is more than the operation limit, the amber DRIVE light is on.

    B. If the IDG oil pressure is less than the operation limit, the amber DRIVE light is on.

    C. When the IDG oil pressure is less than the operation limit, the amber DRIVE light is off.

    D. As if the IDG oil pressure is less than the operation limit, the amber DRIVE light is on.

2. Any one of the power sources can supply the entire electrical network.

    A. None of the power sources can supply the entire electrical network.

    B. Some of the power sources can supply the entire electrical network.

    C. The entire electrical network can be supplied by any one of the power sources.

    D. Any one of the power sources can demand the entire electrical network.

3. Measure the time necessary for the pressure to decrease.

    A. Time the pressure decay.

    B. Count the time necessary for the pressure to decrease.

    C. Record the time for the pressure to go upwards.

    D. Count the time necessary for the pressure to increase.

4. A list of Page and Effectivity should be provided in the front of each Manual.

    A. A list of Page and Effectivity should be printed in the front of each Manual.

    B. A list of Page and Effectivity must be provided in some Manuals.

    C. A list of Page and Effectivity should be provided in the Manual.

    D. A list of Page and Effectivity must be provided in the Manual.

5. In some cases, the Specification is used to prepare for Continuing Airworthiness Program for acceptance by another airworthiness authority other than FAA.

    A. Both FAA and other authority may accept the Specification.

    B. Other authority may accept the Specification, but not FAA.

    C. Either FAA or other authority may accept the Specification.

    D. Neither FAA nor other authority may accept the Specification.

2.3 Change the following components specified in lower case and translate them into Chinese.

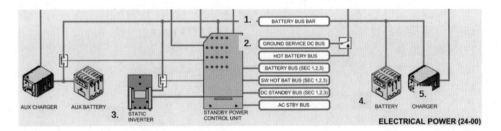

1.	
2.	
3.	
4.	
5.	

Section 3 Aviation Translation: translate the following sentences into Chinese

1. WARNING: PUT A DO-NOT-OPERATE TAG ON THE EXTERNAL POWER RECEPTACLE. IF YOU DO NOT OBEY THIS INSTRUCTION, INJURIES TO PERSONS CAN OCCUR.

2. WARNING: REMOVE ELECTRICAL POWER BEFORE REMOVAL OR INSTALLATION OF THE CURRENT TRANSFORMERS IN THE POWER DISTRIBUTION PANELS. HIGH VOLTAGE PRESENT CAN CAUSE INJURIES TO PERSONS.

11 中国梦·大国工匠篇——航修专家罗卓红：要干就干得最好

Section 4 Aviation Writing

Liu Shuai is an aircraft mechanic. He is inspecting the plane and finds some faults.

Please help him write down fault descriptions. Some related words, phrases & terms and key sentences are offered as follows.

Key words, phrases & terms:

1.	电瓶	battery
2.	外部电源	external power
3.	电压计	voltmeter
4.	变压器	transformer
5.	变频器	converter
6.	整流器	rectifier
7.	保险丝	fuse
8.	起动机	starter
9.	电瓶电压	battery voltage
10.	静变流机	static inverter
11.	备用电源控制组件	standby power control unit
12.	汇流条电源控制组件	bus power control unit
13.	跳开关	circuit breaker
14.	电源接通灯	power ON light
15.	地面勤务电门	ground service switch
16.	变压整流器装置灯	TR UNIT light
17.	转换汇流条	transfer bus
18.	电瓶充电器	battery charger
19.	电流	current
20.	电压	voltage
21.	电阻	resistance
22.	继电器	relay

Key sentences:

1. 地面电源电门在接通位时，灯亮。
   Lights on with GRD PWR switch at ON.
2. APU 发电机接通时，灯亮。
   Lights on with the APU generator on line.

3. 1 号 IDG 和 2 号 IDG 有电时，电源关断和发电机关断汇流条灯亮。
    SOURCE OFF and GEN OFF BUS lights are on with IDG 1 and IDG 2 on line.

4. AC 电源有电且备用电源电门在自动位或电瓶位时，备用电源关断灯亮。
    STANDBY PWR OFF light on with AC power supplied and STANDBY POWER switch
    at AUTO or BAT.

5. 过站检查右发 IDG 加油口漏油。
    The IDG filler cover of R/H engine leaks oil in TR.

6. 更换发电机驱动泵管路滑油滤芯。
    Replace the generator drive line oil filter element.

7. 更换 CSD。
    Replace the CSD.

8. 更换右发 IDG 加油和搜油滤。
    Replace R/H IDG charge and scavenge filters.

Task: Liu Shuai has three writing tasks. The first one is that Number 2 Engine Drive Generator has no output, the second is to supply the external power to the buses, and the last one is to replace the Integrated Drive Generator. Please finish the writing tasks for him.

Warning Case

The Disaster of Continental Express 2574

On September 11, 1991, Continental Express Flight 2574 was about to complete its journey from Laredo to Houston, Texas. When the pilot was about to land, it dived and crashed in the air and its left wing was torn apart. Fourteen people were killed. After investigation, the inspector took part in the work of the maintenance technician and helped to remove the screws on the top of the front of the horizontal stabilizer, because the plan of the day was to replace the deicing devices on both sides. Unexpectedly, the night shift didn't replace the one on the left, and he didn't mention that the screw on the left had been removed during the shift handover. In this way, without the upper screws of the leading edge skin of the left horizontal stabilizer, the aircraft was put into flight the next day, which eventually led to an air crash.

12 警示案例——美国
大陆快运 2574 号班机

Warning Tips: The fundamental reason lies in the poor management of airlines, the lax control of the responsibilities and work processes of various work as well as the failure of maintenance technicians to fill in the handover card.

Lesson 10 External Power

Learning Objectives

1. Knowledge objectives:

   A. To grasp the words, related terms and abbreviations about external power.

   B. To grasp the key sentences about external power.

   C. To know the correct way of application and removal of external power.

2. Competence objectives:

   A. To be able to read and understand frequently-used & complex sentence patterns, capitalized English materials and obtain key information quickly about external power quickly.

   B. To be able to communicate with English speakers about external power.

   C. To be able to fill in job cards in English.

3. Quality objectives:

To be able to self-study with the help of aviation dictionaries, Internet or other resources.

Section 1 Aviation Listening & Speaking

1.1 Aviation Listening: listen to the record and fill in the blanks.

1 听力录音   2 答案及原文

The airplane is 1. _____ with an alternating-current (AC) and a direct-2. _____ (DC) electrical power system. Power for the AC system is 3._____by two alternator-type 4._____.

The DC system normally is powered from the AC system through two 5._____-rectifiers. A 24- volt battery is available for use in an emergency to supply DC power to essential equipment.

During ground operation, AC and DC power can be supplied to the airplane by an 6. _____ power source. During captive flight, the carrier airplane can supply AC and DC power to the airplane. Both external power sources supply 7._____DC power for initial relay or valve operation only. Large amounts of DC power then are supplied from the AC system through the transformer-rectifiers. The external power receptacle is on the upper 8._____of the fuselage, aft of the canopy.

When AC and DC external power is applied to the airplane by a ground 9._____, an 10. _____ must be used. When external power is applied from the carrier airplane, a single plug-in unit in the carrier airplane pylon is used.

1.2 Aviation Speaking: practice the following dialogue and design your own.

Situation:

This is a dialogue between the pilot and the ground mechanic. The ground mechanic fails to restart the APU. He is asked to connect the external power to finish refueling operation and then to check the APU again. Practice it and have your own.

Note: PIL=Pilot, GND=Ground Mechanic.

3 对话录音

Scene 1

PLT: Cockpit to ground. What's going on?

GND: It seems the APU shut down automatically.

PLT: How could this happen?

GND: Maybe it is a protective automatic shutdown on the APU.

PLT: Really? OK. First, we must do a test on APU. If we can't find any existing fault involved in APU starter, we can restart the APU.

GND: Roger. But we got a problem. I still fail to restart the APU.

PLT: If so, connect the external power to finish refueling operation. After refueling, we are going to check the APU again.

GND: Roger.

4 对话译文

Scene 2

GND: Cockpit, the refueling operation completed.

PLT: OK. Please disconnect the fuel hoses from the adapter. I'll try again.

GND: Roger. Fuel hoses disconnected.

PLT: Roger. Start the APU.

GND: APU starts.

PLT: I was wondering if there was an APU fire. Since the APU can be started again, there is no more worry about the APU fire.

Words & Expressions

1. adapter [ə'dæptə] *n.*   适配器，转接头
2. hose [həʊz] *n.*   软管

Section 2 Aviation Reading

Pre-reading questions:

1. When is external power needed?

5 课文朗读   6 课文译文

2. How can you make yourself safe while supplying external power?

External Power Application

External power is the normal source of AC power for the airplane electrical system when the airplane is on the ground. This power source supplies AC power to the airplane. The electrical components change the AC power to DC power. It lets you have the engines and APU power sources off and also gives power to the battery charger to charge the battery.

7 外部电源（微课）

The electrical power for the ground operation of all electrical loads is supplied by an external 3-phase, 115 volt, 400 Hertz alternating current system. The power is supplied through an external power receptacle.

Components and their Locations

The external AC power system consists of the following components and indicating lights:

➢ P 19 external power receptacle.
➢ External Power Contactor (EPC).
➢ Bus Power Control Unit (BPCU) (in the EE compartment).
➢ P 5-13 electrical meters, battery and galley power module (in the flight compartment).
➢ P 5-4 AC systems, generator and APU module (in the flight compartment).

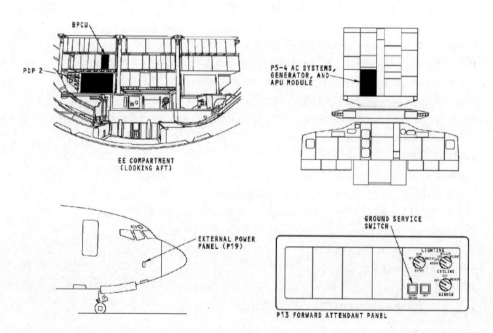

Fig.10-1   Components Locations

The AC external power receptacle is located on the right side, forward of the nose wheel well. The BPCU is on E4 rack. The EPC is in the Power Distribution Panel 2 (PDP 2).

The P 5-4 AC systems, generator, and APU module give you manual control and indication of external power. The ground service switch is on the forward attendant panel.

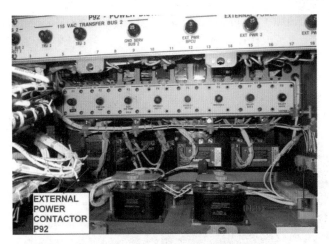

Fig.10-2   External Power Contactor

Fig.10-3   Bus Power Control Unit (BPCU)

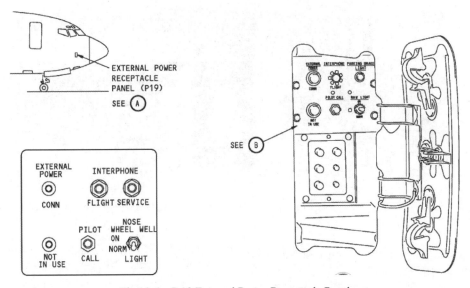

Fig.10-4   P 19 External Power Receptacle Panel

The BAT switch of P 5-13 electrical meters, battery and galley power module lets you energize these buses with battery power:

➤ Switched hot battery bus.

➤ Battery bus.

➤ AC and DC standby buses.

The switch is guarded in the BAT ON position.

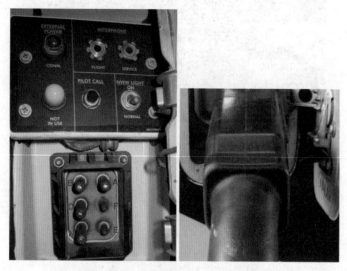

Fig.10-5　P 19 External Power Receptacle & Plug

The ground power switch of P 5-4 AC systems, generator and APU module lets you manually control when good quality ground power is available. The blue GRD POWER AVAILABLE light above the switch is on if the ground power is connected and quality is good. There are three positions for the ground power switch and it is spring loaded to the center position. The ON and OFF positions are momentary positions.

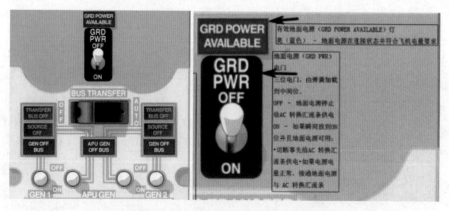

Fig.10-6　GRD POWER AVAILABLE

P 19 external power receptacle panel has two sections: external power receptacle and control and display section. The control and display section has the EXTERNAL POWER CONN and NOT IN USE indicators.

The amber EXTERNAL POWER CONN light comes on when the ground power plug is connected and power is present. The white NOT IN USE indicator comes on when external power is present but not in use.

The external power source connects to the airplane external power receptacle. The Bus Protection Panel (BPP) is used for control and protection of the airplane circuits connected to external power. The two external power contactors connect 115 volt AC, 400 Hertz, 3-phase power to the distribution system from external power source.

Manual Control

The ground power switch is on the P 5-4 AC systems, generator, and APU module. Use this switch to control external power to the AC transfer buses through the EPC and the BTBs. The BPCU closes the EPC directly. Each BTB closes after it receives a signal from the BPCU through a GCU.

A switch on the forward attendant panel lets you supply power to the AC and DC ground service buses from the external power source. The AC power goes through the two ground service transfer relays. The DC power goes through the ground service bus relay. The BPCU uses input from the forward attendant panel to control the relays.

External Power Supply and Removal

External Power Supply

Operate the external power supply correctly while supplying external power to the airplane. First, open the external power receptacle door and connect the power cable to P 19 external power receptacle. Then energize the external power cable.

The following lights show:

➢ EXTERNAL POWER CONN(amber)(on P 19 external power panel).

➢ NOT IN USE(white) (on P 19 external power panel).

➢ GRD POWER AVAILABLE light (blue) (on P 5-4 panel).

Fig.10-7　P 5-4 Panel

Let the ground power go to the AC transfer buses, you should do the following steps. Put the GRD POWER AVAILABLE switch (on P 5-4 panel) to ON. The lights on the P 5-4 panel go off when ground power goes to the AC transfer buses:

➢ 1 source off.

➢ 2 source off.

➢ 1 transfer bus off.

➢ 2 transfer bus off.

The not in use light (on the P 19 panel) also goes off.

WARNING: IF THE EXTERNAL POWER SUPPLY HAS AN EARTH GROUNDED NEUTRAL, THERE MUST NOT BE AN OPEN OR FLOATING GROUND IN THE NEUTRAL CIRCUIT WIRING OF THE SUPPLY OR THE AIRPLANE. IF AN OPEN OR FLOATING GROUND IS PRESENT, THE AIRPLANE CAN BE PUT AT AN ELECTRICAL POTENTIAL ABOVE EARTH GROUND. THIS ELECTRICAL POTENTIAL CAN CAUSE ELECTRIC SHOCK WITH POSSIBLE SEVERE INJURIES TO PERSONNEL WHO TOUCH THE AIRPLANE.

WARNING: REMOVE THE ELECTRICAL POWER FROM THE CABLE BEFORE CONNECTING THE CABLE TO THE AIRPLANE. INJURIES TO PERSONS CAN BE CAUSED BY AN ELECTRICAL SHOCK.

External Power Removal

To remove external power, do the following steps.

First, put the GRD PWR AVAILABLE switch (P 5-4) momentarily to the OFF position. Then put the BAT switch (P 5-13) to the OFF position. This removes the AC power from the AC transfer buses, and the battery power from the standby, switched hot battery and battery buses. The NOT IN USE light (P 19) illuminates again. When de-energizing the external power cable, the EXTERNAL POWER CONN and the NOT IN USE light go off. Now the cable can be safely removed from the P19 external power receptacle.

Fig.10-8   External Power Removal

Abbreviations & Acronyms

1.	External Power (EP)	外部电源
2.	External Power Contactor (EPC)	外部电源接头
3.	Bus Power Control Unit (BPCU)	汇流条功率控制组件
4.	Auxiliary Power Unit (APU)	辅助动力装置
5.	Ground Power (GRD PWR)	地面电源
6.	External Power Contactor (EPC)	外部电源接触器
7.	Bus Protection Panel (BPP)	汇流条保护板
8.	Bus Tie Breaker (BTB)	汇流条断路器
9.	Generator Control Unit (GCU)	发电机控制组件

Words & Expressions

1.	battery charger	电瓶充电器

2. charge [tʃɑːdʒ] *v.*　　　　给……充电

3. phase [feɪz] *n.*　　　　相

4. volt [vəʊlt] *n.*　　　　伏特

5. Hertz [hɜːts] *n.*　　　　赫兹

6. receptacle [rɪˈseptəkl] *n.*　　　　插板

7. contactor [ˈkɒntæktə] *n.*　　　　接触器，接头

8. electrical meter　　　　电表

9. galley [ˈgæli] *n.*　　　　机上厨房

10. generator [ˈdʒenəreɪtə(r)] *n.*　　　　发电机

11. rack [ræk] *n.*　　　　支架

12. switch [swɪtʃ] *n.*　　　　电门，开关

13. panel [ˈpænl] *n.*　　　　面板

14. distribution [ˌdɪstrɪˈbjuːʃn] *n.*　　　　分配

15. attendant [əˈtendənt] *n.*　　　　服务员

16. energize [ˈenədʒaɪz] *v.*　　　　充电

17. standby [ˈstændbaɪ] *adj.*　　　　备用的

18. guard [gɑːd] *v.*　　　　保护，防护

19. ground power　　　　地面电源

20. available [əˈveɪləbl] *adj.*　　　　可获得的

21. momentary [ˈməʊməntri] *adj.*　　　　暂时的

22. indicator [ˈɪndɪkeɪtə(r)] *n.*　　　　指示器

23. amber [ˈæmbə(r)] *adj.*　　　　琥珀色的

24. plug [plʌg] *n.*　　　　插座

25. source [sɔːs] *n.*　　　　电源

26. relay [ˈriːleɪ] *n.*　　　　继电器

27. go off　　　　停止运转

28. neutral [ˈnjuːtrəl] *adj.*　　　　中间的

29. floating [ˈfləʊtɪŋ] *adj.*　　　　浮动的

30. circuit [ˈsɜːkɪt] *n.*　　　　电路

31. potential [pəˈtenʃl] *n.*　　　　电势

32. electric shock　　　　电击

33. severe [sɪˈvɪə(r)] *adj.*　　　　十分严重的

34. removal [rɪˈmuːvl] *n.*　　　　移除

35. transfer [trænsˈfɜː(r)] *v.*　　　　转换

36. illuminate [ɪˈluːmɪneɪt] *v.*　　　　照亮

37. de-energize [diːˈenədʒaɪz] *v.*　　　　断电

38. cable [ˈkeɪbl] *n.*　　　　电缆

8 单词朗读

9 术语专讲：计量表
（微课）

Exercises

2.1 Choose the best answer from A, B, C, D options.

1. rack

　　A. 千斤顶　　　　B. 支架　　　　C. 开关　　　　D. 碎片

2. source

　　A. 力量　　　　B. 过程　　　　C. 资源　　　　D. 电源

10 答案

3. indicator
    A. 象征　　　　　B. 指示　　　　　C. 指示器　　　　　D. 指示者
4. circuit
    A. 电路　　　　　B. 圆圈　　　　　C. 循环　　　　　D. 环绕
5. de-energize
    A. 通电　　　　　B. 放电　　　　　C. 无力　　　　　D. 精力旺盛的
6. 琥珀色的
    A. ambitious　　　B. ambulance　　　C. ampere　　　　D. amber
7. 中间的
    A. naughty　　　　B. neutral　　　　C. nervous　　　　D. neutralize
8. 电表
    A. electrical meter　B. electrical power　C. electrical shock　D. electric resistance
9. 通电测试
    A. power-off test　B. pressure test　　C. power-on test　　D. friction test
10. 外部电源
    A. internal power　B. external power　C. ground power　　D. electrical power

2.2 Find the best option to paraphrase the sentence.

1. Some components must be well kept for normal use.
    A. Some components can not be used normally if they are well kept.
    B. Some components must be well kept so that they can be used normally.
    C. Some components must be well kept as long as they can be used normally.
    D. Some components could be well kept so that they can be used abnormally.
2. Each Chapter will start with a new block of page numbers.
    A. Each Chapter will use a new block of page numbers as a beginning.
    B. When you prepare each Chapter, you can continue to use the page numbers previously-used.
    C. The beginning of each Chapter shall list a block of page numbers previously-used.
    D. Most Chapters will start with a new block of page numbers.
3. During an in-flight start, the EEC selects the two ignition exciters to get power.
    A. During an in-flight start, the EEC selects either ignition exciter to get power.
    B. Two ignition exciters are chose by EEC to get power during an in-flight start.
    C. During an in-flight start, one ignition exciter is active and the other one standby.
    D. During an in-flight start, the EEC selects neither ignition exciter to get power.
4. Remove the electrical power from the cable before connecting the cable to the airplane.
    A. Replace the electrical power from the cable before connecting the cable to the airplane.
    B. Remove the electrical power from the cable when connecting the cable to the airplane.
    C. Remove the electrical power from the cable after connecting the cable to the airplane.
    D. Eliminate the electrical power from the cable before connecting the cable to the airplane.
5. This electric potential can cause electric shock with possible sever injuries to personnel who touch the airplane.
    A. This electric potential can effect electric shock with possible sever injuries to personnel who touch the airplane.
    B. This electric potential can result in electric shock with possible sever injuries to personnel who touch the airplane.

    C. This electric potential can result from electric shock with possible sever injuries to personnel who touch the airplane.

    D. This electric potential can cause electric shock with possible sever destruction to personnel who touch the airplane.

2.3 Change the following components specified in lower case and translate them into Chinese.

1.	
2.	
3.	
4.	
5.	

Section 3 Aviation Translation: translate the following sentences into Chinese

1. WARNING: REMOVE THE ELECTRICAL POWER FROM THE EXTERNAL POWER CABLE BEFORE YOU CONNECT THE CABLE TO THE AIRPLANE. INJURIES TO PERSONS CAN BE CAUSED BY AN ELECTRICAL SHOCK.

2. CAUTION: WHEN YOU CLOSE THE BATTERY SWITCH, THE BATTERY AND STANDBY BUSES HAVE POWER. YOU WILL DISCHARGE THE BATTERY TO ZERO VOLTS VERY FAST. YOU SHOULD GET AN AC POWER SOURCE ON THE ELECTRICAL SYSTEM QUICKLY.

11 坚守岗位，不负使命
——海航集团金鹏航空
维修技术员罗湘军用
脚步践行（文本）

Section 4 Aviation Writing

Liu Shuai is an aircraft mechanic. He is inspecting the plane and finds some faults.

Please help him write down fault descriptions. Some related words, phrases & terms and key sentences are offered as follows.

Key words, phrases & terms:

1. 电源测试   PWR TEST
2. 通电测试   power-on test
3. 在空中   in flight/ in the air
4. 地面检查   GND check
5. 外部电源   external power
6. 电压计   voltmeter
7. 勤务汇流条   service bus
8. 转换汇流条   transfer bus
9. 电瓶充电器   battery charger
10. 电流   current
11. 电压   voltage
12. 电瓶电压   battery voltage
13. 汇流条功率控制组件   bus power control unit
14. 跳开关面板灯   circuit breaker panel light
15. 开路   open circuit

Key sentences of direction description:

1. 给汇流条提供外部电源。

   Supply the external power to the buses.
2. 地面电源电门在接通位时，灯亮。

   Lights on with GRD PWR switch at ON.
3. APU 发电机接通时，灯亮。

   Lights on with the APU generator on line.
4. 12B 前排座椅下的线盒破损，电源外露。

   The terminal block under the forward Seat 12B is damaged and the electric is naked.

Task: Liu Shuai has three writing tasks. The first one is to jack at external power receptacle panel, the second is that TRANSFER BUS OFF and SOURCE OFF lights is off, and the last one is that SOURCE OFF and GEN OFF BUS lights come on with IDG 1 & 2 on line. Please finish the writing tasks.

Warning Case

Chalk's Ocean Airways 101- Right Wing Fractured

12 警示案例——卓克斯海洋航空 101 号班机—右翼断裂

On December 19, 2005, Chalk's Ocean Airways 101 took off from Miami, Florida. A few seconds later, the right wing of the aircraft broke due to metal fatigue, causing the aircraft to crash in the sea, killing 20 people.

Warning Tips: the probable cause of this accident was the in-flight failure and separation of the right wing during normal fight, which resulted from the next two reasons. First, Chalk's marine Airlines maintenance plan failed to identify and properly repair the fatigue cracks of the right wing. Second, FAA failed to identify and correct defects in the company's maintenance plan.

Lesson 11 Fire Protection

Learning Objectives

1. Knowledge objectives:

   A. To grasp the words, related terms and abbreviations about fire protection system.

   B. To grasp the key sentences about fire protection system.

   C. To know the importance of fire protection system on the aircraft.

2. Competence objectives:

   A. To be able to read and understand frequently-used & complex sentence patterns, capitalized English materials and obtain key information about fire protection system quickly.

   B. To be able to communicate with English speakers about the topic freely.

   C. To be able to fill in the job cards in English.

3. Quality objectives:

   A. To be able to self-study with the help of aviation dictionaries, the Internet and other resources.

   B. To develop the craftsman spirit of carefulness and responsibility.

Section 1 Aviation Listening & Speaking

1.1 Aviation Listening: listen to the record and fill in the blanks.

1 听力录音　　2 答案及原文

In this lesson, we will discuss the fire 1.＿＿＿＿＿system fitted to all aircraft. A fire cannot exist without three things being present. These are heat or a source of ignition bill material that will 2.＿＿＿＿＿and oxygen. This is known as the triangle of fire. Take any one of the three away and the fire will go out. A fire in an aircraft is extremely 3.＿＿＿＿＿. So to 4.＿＿＿＿＿the aircraft, the crew and the passengers, aircraft must have fire detection and 5.＿＿＿＿＿systems fitted in all areas where a potential fire risk may exist following failure of or leakage from any component or associated equipment, and some specific areas also require protection, including the 6.＿＿＿＿＿, the auxiliary power unit, or a put the main wheel wells, and the cargo compartments. These areas are known as designated fire 7._____, and they must have detection systems fitted to 8._____the crew of a fire. In the case of the engines and the auxiliary power unit, they are 9._____by fireproof bulkheads, known as fire walls, normally manufactured from titanium or stainless steel, which will contain any fire to the immediate area and prevent it from 10._____ to the rest of the aircraft.

1.2 Viation Speaking: practice the following dialogue and design your own.

Situation:

This is a dialogue between two ground mechanics about performing the engine running rest after troubleshooting of fuel leakage. Practice it and have your own.

3 对话录音

Note: GND=Ground Mechanic

Scene 1

GND A: The fuel control unit has been installed. Now we can perform the engine running test.

GND B: Roger. Ready to test the engine. Wait a minute. The velocity of wind is too high, we need to tow the aircraft and make it facing the wind direction so that we can test the engine.

4 对话译文

GND A: Roger. Copy that. Tractor will be ready in 5 minutes.

Scene 2

GND A: Hello. Tractor is ready.

GND B: Roger. Please tow the aircraft and make it facing the wind direction.

GND A: Roger, copy that, facing the wind direction. Sir, we are in the right position.

GND B: Roger, wait to start the engine.

GND A: Roger. Disconnect the tow-bar. Please set the parking brake.

GND B: Parking brake is set. Make sure the ground is clear and fire equipment is in position.

GND A: The ground is clear. Fire fighter stands by. Ground is ready.

GND B: Copy that, we'll start engine one.

Words & Expressions

1. perform [pə'fɔːm] *v.*   执行
2. velocity [və'lɒsəti] *n.*   速度

Section 2 Aviation Reading

Pre-reading questions:

1. What should we do in case of fire on the aircraft?
2. Why is fire protection needed?

5 课文朗读　　6 课文译文　　7 防火（微课）

Fire Protection

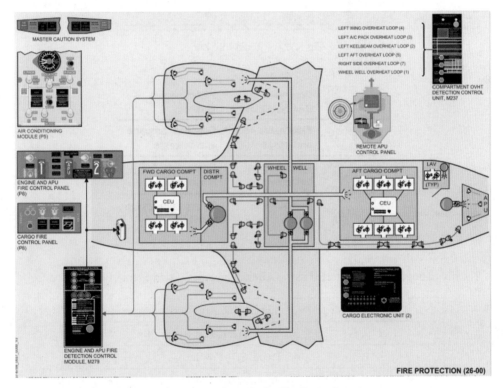

Fig.11-1　Schematic for Fire Protection

Fire protection is the study and practice of mitigating the unwanted effects of potentially destructive fires. It involves the study of the behavior, compartmentalisation, suppression and investigation of fire and it's related emergencies, as well as the research and development, production, testing and application of mitigating systems. Let's take the example of the fire protection system on the aircraft to illustrate.

Generally the aircraft fire zones on an aircraft include cockpit, cabin, cargo compartments, galleys and toilets, wheel wells, leading edges, and so on.

Fig.11-2　Aircraft Fire Zones

A		Common Combustibles	Wood, paper, cloth etc.
B		Flammable liquids and gases	Gasoline, propane and solvents
C		Live electrical equipment	Computers, fax machines (see note!)
D		Combustible metals	Magnesium, lithium, titanium

Fig.11-3　Fire Classification on the Aircraft

The fire extinguishing agent used in the Aircraft includes water, Freon, Powder, and CO_2.

Fire Classification	Suitable Fire Extinguishing Agent
A	Water, Freon, Powder
B	Freon, CO_2, Powder
C	Freon, CO_2, Powder
D	Powder

Fig.11-4　Fire Extinguishing Agent Used in the Aircraft

According to FAR 25, Engine nacelles, APU compartments and Fuel burning heaters must be provided with fire detection systems and built in fire extinguisher systems. Hand fire extinguishers are necessary in the cockpit, cabin, class A and B cargo compartment.

The fire protection system monitors the fires, smoke, overheat and pneumatic duct leaks on the airplane. A complete fire protection system on modern aircraft includes fire detection, overheat detection, fire extinguishing system and so on.

> ➢ Fire detection system.
> ➢ Overheat detection system.
> ➢ Smoke detection system.
> ➢ Fire extinguishing system.
> ➢ Lavatory fire extinguishing system.
> ➢ Control and indication of built in fire extinguisher system.

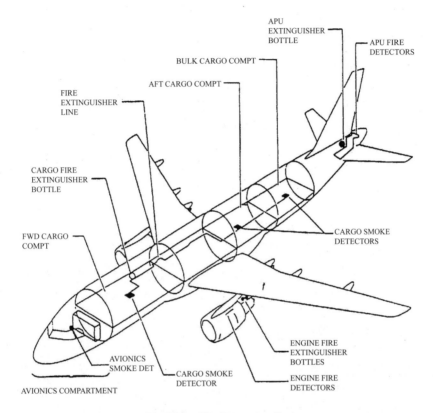

Fig.11-5   Fire Protection System

Fire Detection System

Detection Sensors

Three types of sensors and fire detection system are:

> ➢ Wire type sensor (electrical sensor element).
> ➢ Gaseous sensor/ responder tubes.
> ➢ Infrared ray sensor.

Detection Methods

The detection methods include those most commonly used in turbine engine aircraft fire protection systems. The complete aircraft fire protection system of most large turbine engine aircraft will incorporate several of these different detection methods.

> ➢ Rate-of-temperature-rise detectors.

> Radiation sensing detectors.
> Smoke detectors.
> Overheat detectors.
> Carbon monoxide detectors.
> Combustible mixture detectors.
> Fiber optic detectors.
> Observation of crew or passengers.

The rate-of-temperature-rise, radiation sensing and overheat detectors are the most commonly used for fast detection of fires.

Detectors' Locations

To detect fires or overheat conditions, detectors are placed in the various zones to be monitored. The fire/overheat detection system includes: engine overheat detection, engine fire detection, APU fire detection, wheel well fire detection, wing/body overheat detection, cargo compartment smoke detection and lavatory smoke detection.

The following components on the airplane have fire extinguishing system: the engine, APU, lavatory, cargo compartment and portable fire extinguishers.

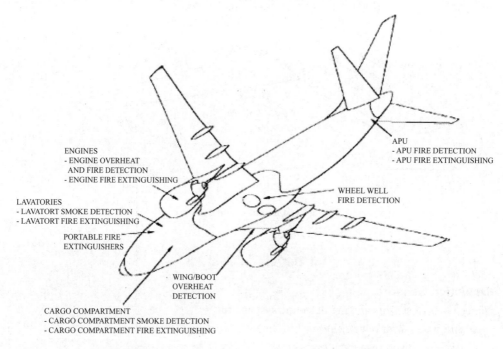

Fig.11-6   Detectors' Locations of Fire Protection System

Fire Extinguishing System

According to FAR 25, one fire extinguisher is required for APU. Two fire extinguishers are required for any main engine.

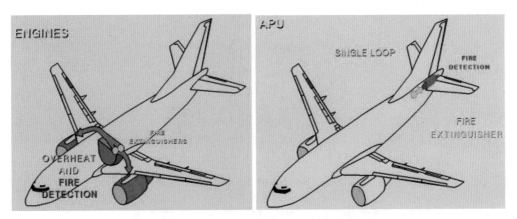

Fig.11-7   Locations of Fire Extinguishing System

Built in fire extinguishing system includes:

➢ Fixed fire extinguisher uses freon gas as the extinguishing agent with nitrogen gas used to pressurize the bottles.

➢ Overpressure indicator (red).

➢ Discharge indicator (yellow).

➢ Pulling out the fire switch/ handle.

➢ Preparing for discharging the extinguishing agent.

➢ Turning the fire switch/ handle to discharge the extinguishing agent, bottle discharged lights on.

Freon gas has the following extinguishing effects:

➢ Anticatalytic effect: Lower the velocity of the chemical reaction (burning).

➢ Cooling effect: Expansion of freon.

➢ Suffocation effect: Supplant the oxygen since freon is heavier than air.

WARNING: IN ORDER TO PREVENT POISONING, AVOID CONTACT WITH THE AGENT, DO NOT BREATHE THE AGENT IN VENTILATE AREAS BEFORE ENTERING.

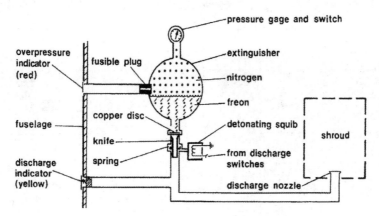

Fig.11-8   Fixed Fire Extinguisher

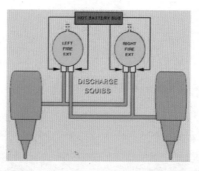

Fig.11-9　Overpressure & Discharge Indicator

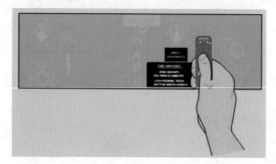

Fig.11-10　Pulling out the Fire Switch/ Handle

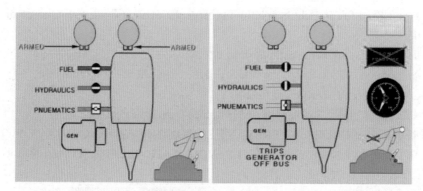

Fig11-11　Preparing for Discharging the Extinguishing Agent

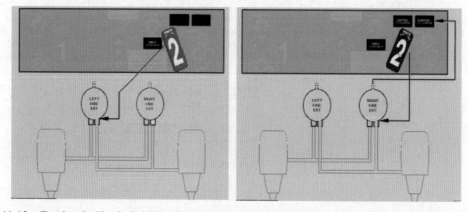

Fig.11-12　Turning the Fire Switch/ Handle to Discharge the Extinguishing Agent, Bottle Discharged Lights on

Different Types of Fire Extinguisher

There are portable fire extinguishers to extinguish fires inside the airplane. There are different types of extinguishers. There might be water-type, carbon dioxide or dry chemical fire extinguisher. The water-type extinguishers are for solid combustible fires. The Freon/halon extinguishers are for electrical and flammable liquid fires. Freon/halon fire extinguishers are usually used in aircraft.

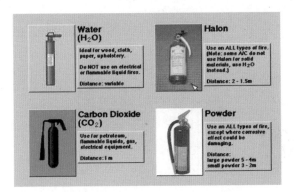 

Fig.11-13   Different Types of Fire Extinguishers   Fig.11-14   Water Fire Extinguisher

The water-type extinguishers contain a water and anti-freeze mixture. Each water extinguisher has these parts:

> Trigger (discharge valve).
> Water cylinder.
> Handle with internal gas cartridge.
> Discharge nozzle.
> Quick-release mounting strap.

Each Halon extinguisher has these parts:

> Pressure indicator.
> Halon cylinder.
> Trigger.
> Discharge nozzle.
> Handle lock pin.
> Handle.

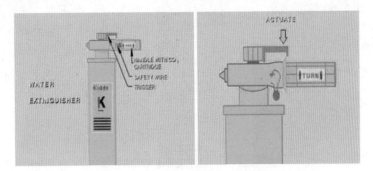

Fig.11-15   A Water-Type Extinguisher

Fig.11-16　Hand Fire Extinguisher

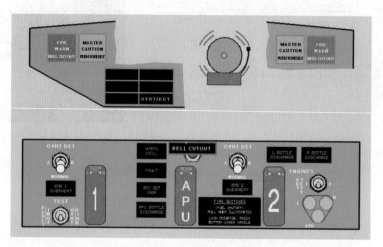

Fig.11-17　Control & Indication of Fire Extinguisher System

Fire Alarms

The fire alarms provide visual and aural indication to the flight crew about a fire in any of the following components:

➢ Engine 1& 2.
➢ APU.
➢ Forward & aft cargo compartment.
➢ The main wheel well.

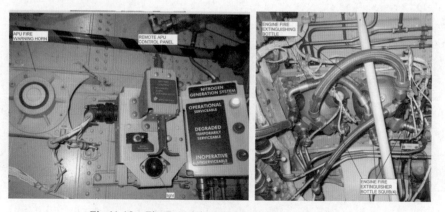

Fig.11-18　Fire Protection Components on the Aircraft

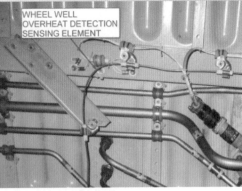

Fig.11-19　Fire Protection Components on the Aircraft（续图）

Visual & Aural Indications

When there is a fire, these are the visual and aural indications shown:

➢ Two red FIRE WARN lights on the P 7 panel come on.

➢ Bell in the aural warning unit comes on.

➢ Red light and horn in the right main wheel well come on alternately (APU fire only). The horn does not come on in flight.

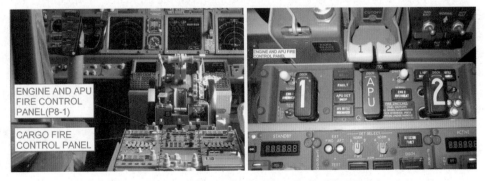

Fig.11-20　Fire Protection Components on the Aircraft

When there is a fire in the engine, the engine and APU fire detection module provides a ground for the two red FIRE WARN lights and the bell.

When there is a fire in the APU, the engine and APU fire detection module provides a ground for the two red FIRE WARN lights and the bell. The same module provides power for the red light and horn on the P 28 panel.

When there is a fire in the main wheel well, the compartment overheat detection controller provides a ground for the two red FIRE WARN lights and the bell.

When there is smoke in the forward or aft cargo compartment, the cargo electronic unit provides a ground for the two red FIRE WARN lights and the bell.

The two red FIRE WARN lights, bell, and horn are reset by removal of the ground. The reset is done by a momentary push on left FIRE WARN light, right FIRE WARN light, bell cutout switch on P 8 or horn cutout switch on the P 28 panel.

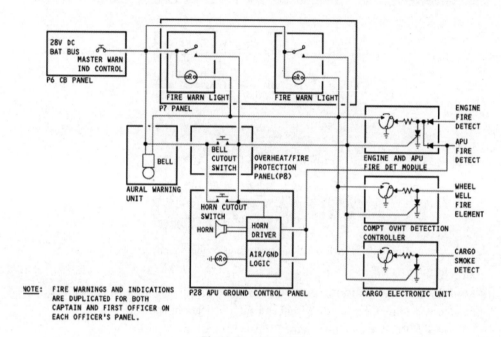

Fig.11-21　Fire Alarms

Fire Prevention and Protection

Leaking fuel and hydraulic, de-icing or lubricating fluids can be sources of fire in the aircraft. When inspecting aircraft systems, correct action should be taken immediately. Minute pressure leaks of these fluids are particularly dangerous for they quickly produce an explosive atmospheric condition. Carefully inspect fuel tank installations for signs of external leaks. With integral fuel tanks the external evidence may occur at some distance from where the fuel is actually escaping.

Many hydraulic fluids are flammable and should not be permitted to accumulate in the structure.

Sound proofing and lagging materials may become highly flammable if soaked with oil if any kind.

Any leakage or spillage of flammable fluid in the vicinity of combustion heaters is a serious fire risk, particularly if any vapor is drawn into the heater and passes over the hot combustion chamber.

Oxygen system equipment must be kept absolutely free from traces of oil or grease, since these substances will spontaneously ignite in contact with oxygen under pressure. Oxygen servicing cylinders should be clearly marked so that they cannot be mistaken for cylinders containing air or nitrogen, as explosions have resulted from this error during maintenance operations. Fire prevention is much more rewarding than fire extinguishing.

(a) Components in the Flight Compartment

(b) Components in the WW

(c) Components in the Cargo Compartment

Fig.11-22 Fire Protection Components on the Aircraft

Abbreviations & Acronyms

1.	Extinguish (EXT)	熄灭
2.	Overheat (OVHT)	过热
3.	Detector (DET)	探测器
4.	Valve (VLV)	活门

Words & Expressions

1. mitigate ['mɪtɪgeɪt] *v.*   减轻
2. compartmentalization [kɒmpɑːtmentəlaɪ'zeɪʃn] *n.*   划分
3. suppression [sə'preʃn] *n.*   抑制
4. fire detection   火警探测
5. fire protection system   防火系统
6. fire extinguishing   灭火
7. extinguisher [ɪk'stɪŋgwɪʃə(r)] *n.*   灭火器
8. nacelle [nə'sel] *n.*   引擎舱，吊舱
9. freon ['friːɒn] n   氟利昂
10. heater ['hiːtə(r)] *n.*   加热器
11. pneumatic [njuː'mætɪk] *adj.*   气动的
12. duct [dʌkt] *n.*   管道；输送管
13. lavatory ['lævətri] *n.*   厕所
14. sensor ['sensə(r)] *n.*   传感器

8 单词朗读

9 术语专讲：管道
（微课）

15. gaseous ['gæsiəs] *adj.*   气态的

16. infrared [ˌɪnfrəˈred] *adj.*   红外线的

17. turbine engine   涡轮发动机

18. radiation sensing detector   辐射传感探测器

19. carbon monoxide   一氧化碳

20. combustible mixture   可燃混合物

21. fiber optic   光纤

22. portable ['pɔːtəbl] *adj.*   便携的

23. nitrogen ['naɪtrədʒən] *n.*   氮

24. anticatalytic [æntɪkætəˈlɪtɪk] *adj.*   反催化的

25. velocity [vəˈlɒsəti] *n.*   速度

26. expansion [ɪkˈspænʃn] *n.*   膨胀

27. suffocation effect   火焰窒息作用

28. halon ['heɪlɒn] *n.*   卤化烷

29. trigger ['trɪgə(r)] *n.*   触发器

30. cylinder ['sɪlɪndə(r)] *n.*   储氧筒

31. cartridge ['kɑːtrɪdʒ] *n.*   弹药筒；暗盒

32. nozzle ['nɒzl] *n.*   喷嘴，喷管

33. strap [stræp] *n.*   胶带，带子

34. pin [pɪn] *n.*   销子

35. visual ['vɪʒuəl] *adj.*   视觉的

36. aural ['ɔːrəl] *adj.*   听觉的

37. aural warning unit   音频警告组件

38. horn [hɔːn] *n.*   喇叭

39. cutout switch   切断电门

40. installation [ˌɪnstəˈleɪʃn] *n.*   安装

41. integral ['ɪntɪgrəl] *adj.*   完整的

42. sound proofing   隔音

43. lagging material   绝缘材料

44. soak [səʊk] *n.*   浸泡，吸入

45. spillage ['spɪlɪdʒ] *n.*   溢出

46. in the vicinity of   在……附近

47. vapor ['veɪpə(r)] *n.*   蒸汽

48. spontaneously [spɒnˈteɪnɪəsli] *adv.*   自然地，自发地

49. ignite [ɪgˈnaɪt] *v.*   点燃

Exercises

2.1 Choose the best answer from A, B, C, D options.

1. duct

  A. 管理   B. 管道   C. 生产   D. 产品

2. fiber optic

  A. 纤维   B. 光纤   C. 电缆   D. 亚麻

3. extinguish

  A. 熄灭   B. 辨别   C. 灭绝   D. 熄火器

10答案

4. portable

    A. 港口　　　　B. 一部分　　　　C. 比例　　　　D. 便携的

5. carbon monoxide

    A. 一氧化氮　　B. 二氧化氮　　　C. 一氧化碳　　D. 二氧化碳

6. 探测器

    A. detect　　　　B. detector　　　C. detective　　　D. detection

7. 视觉的

    A. visual　　　　B. aural　　　　C. vision　　　　D. visualize

8. 氮

    A. carbon　　　　B. oxygen　　　C. nitrogen　　　D. hydrogen

9. 可燃混合物

    A. combustor　　　　　　　　B. combust mixture

    C. combustion mixture　　　　D. combustible mixture

10. 过热探测器

    A. overhead detector　　　　　B. overheat detector

    C. overhear detector　　　　　D. overheard detector

2.2 Find the best option to paraphrase the sentence.

1. To detect fires or overheat conditions, detectors are placed in the various zones to be monitored.

    A. To protect fires or overheat conditions, detectors are placed in the various zones to be monitored.

    B. To detect fires or overheat conditions, detectors are displaced in the various zones to be monitored.

    C. To detect fires or overheat conditions, detectors are removed in the various zones to be monitored.

    D. To detect fires or overheat conditions, detectors are assigned to the various zones to be monitored.

2. Two fire extinguishers are required for any main engine.

    A. Two fire extinguishers are required for no main engine.

    B. Two fire extinguishers are responsible for any main engine.

    C. Two fire extinguishers are necessary for any main engine.

    D. Two fire extinguishers are unnecessary for any main engine.

3. The water-type extinguishers contain a water and anti-freeze mixture.

    A. The water-type extinguishers consist of a water and anti-freeze mixture.

    B. The water-type extinguishers exclude a water and anti-freeze mixture.

    C. The water-type extinguishers maintain a water and anti-freeze mixture.

    D. The water-type extinguishers obtain a water and anti-freeze mixture.

4. Many hydraulic fluids are flammable and should not be permitted to accumulate in the structure.

    A. Many hydraulic fluids are flammable and should be allowed to accumulate in the structure.

    B. Many hydraulic fluids are flammable and must be permitted to accumulate in the structure.

    C. Many hydraulic fluids are flammable and should be forbidden accumulating in the structure.

D. Many hydraulic fluids are flammable and could not be permitted to deplete in the structure.

5. If you accidentally move the lever, set the system to the neutral position again.

A. If you absolutely move the lever, set the system to the left position again.

B. If you unintentionally move the lever, set the system to the neutral position again.

C. If you accurately move the lever, set the system to the right position again.

D. If you deliberately move the lever, set the system to the center position again.

2.3 Change the following components specified in lower case and translate them into Chinese.

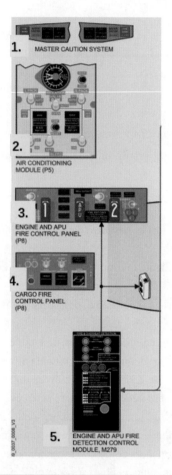

1.	
2.	
3.	
4.	
5.	

Section 3 Aviation Translation: translate the following sentences into Chinese

11 中国制造——
鲲龙（文本）

1. NOTE: IGNORE THE INDICATION LIGHTS FOR THE APU DETECTION SYSTEM AND THE INDICATIONS ON THE ENGINE AND APU FIRE DETECTION CONTROL MODULE, M279, IN THE ELECTRONIC EQUIPMENT BAY.

2. WARNING: DO NOT TOUCH THE LIGHT WITH YOUR BARE HANDS UNTIL THE LIGHT IS OFF FOR FIVE MINUTES. DURING THIS TIME, YOU WILL INJURE IF YOU TOUCH THE LIGHT. THE LIGHT CAN BURN YOUR SKIN OR GIVE YOU AN ELECTRICAL SHOCK.

Section 4 Aviation Writing

Liu Shuai is an aircraft mechanic. He is inspecting the plane and finds some faults.

Please help him write down fault descriptions. Some related words, phrases & terms and key sentences are offered as follows.

Key words, phrases & terms:

1. 灭火瓶   fire extinguisher bottle
2. 火警灯   fire warning light
3. 火警，假警报   fire alarm, false alarm
4. 发动机支架   engine strut
5. 测试电门   test switch
6. 灭火测试电门   extinguisher switch
7. 烟雾探测   smoke detector
8. 头顶探测器   overhead detector
9. 过热探测器   overheat detector
10. 大翼过热探测器   wing overheat detector
11. 厕所烟雾指示灯   lavatory smoke indicator light
12. 过热探测控制组件   overheat detector control module
13. 爆炸帽   squib
14. 火警探测环路   fire detection loop
15. 火警铃   warning bell
16. 失效监控电路   fault monitoring circuit
17. 热电门   thermal switch
18. 手提灭火瓶   portable fire extinguisher
19. 翼身管道过热探测   wing/body duct leak detection
20. 厕所烟雾探测   lavatory smoke detection
21. 货舱烟雾探测   cargo smoke detection
22. 封口破损或丢失   seal broken or missing
23. 客舱有烟雾或气体   fumes or smoke in cabin

Key sentences:

1. 故障 / 不工作和过热 / 火测试电门。
   FAULT/INOP and OVHT/FIRE test switch.

2. APU/ 发动机左 / 右灭火瓶已释放。

167

APU/Engine left/right bottle discharged.

3. 发动机运转时灭火手柄在拉出位。

Fire handle is in the pulled position while engine turns.

4. 机舱内出现烟雾或气体。

Smoke or fumes found in cabin.

5. APU 排气中出现烟雾或火焰。

Smoke or flame found in APU exhaust.

6. 再循环风扇关断，隔离活门关闭且两个组件电门在关断位时，烟雾不停止。

Smoke does not stop with recirculation fans off, isolation valve closed, and both pack switches off.

7. 再循环风扇关断时，烟雾停止。

Smoke stops with recirculation fan(s) off.

8. 再循环风扇关断，隔离活门关闭，且右组件电门在关断位时，烟雾停止。

Smoke stops with recirculation fan(s) off, isolation valve closed, and right pack switch off.

9. 货舱火警探测。

Cargo fire detection.

10. 前 / 后货舱火警灯亮（假火警）。

FWD/AFT cargo fire light on (false alarm).

11. 按压释放电门时，灭火瓶不释放灭火剂。

Bottle does not release extinguishant when DISCH switch is pushed.

12. 按 / 未按压释放电门时，释放灯不亮。

DISC light is not on when DISCH switch is pushed/not pushed.

Task: Liu Shuai has three writing tasks. The first one is that there is something wrong with extinguishing test, the second is that there are momentary flames from APU exhaust when main engine start, and the last one is that the crew orally reported that the fire warning system of the aft cargo compartment did not work. Please finish the writing tasks for him.

Warning Case

Swissair 111 – Fire on Board

12 警示案例——瑞士
航空 111—机舱烈焰

On September 2nd, 1998, during the flight of Swissair 111, there was a sudden smell of smoke in the cockpit. A spark caused by a short circuit in the wire ignited the PET insulation causing the fire. As a result, the electronic components on the plane failed one by one. The flight crashed into the Atlantic Ocean and disintegrated, leaving all 229 people on board dead.

Warning Tips: the cause of the fire is the first-class entertainment system. Adding such a device to the normal circuit system of the aircraft undoubtedly increased the burden of the system. What is more fatal is that such a system was not equipped with a switch to turn it off when necessary. Finally, the overheated wire ignited the insulation layer and led to a fire.

Lesson 12 Lighting System

Learning Objectives

1. Knowledge objectives:

   A. To grasp the words, related terms and abbreviations about lighting system on the aircraft.

   B. To grasp the key sentences about lighting system.

   C. To know the types and functions of lighting system.

2. Competence objectives:

   A. To be able to read and understand difficult English sentences, capitalized aviation English materials and obtain key information about lighting system.

   B. To be able to communicate with the foreign speakers freely.

   C. To be able to fill in the job cards in English.

3. Quality objectives:

   A. To be able to self-study with the help of aviation dictionaries, the Internet and other resources.

   B. To develop the craftsman spirit of carefulness and responsibility.

Section 1 Aviation Listening & Speaking

1.1 Aviation listening: listen to the record and fill in the blanks.

1 听力录音   2 答案及原文

There are eight different types of exterior lights. They are 1._____ light, take-off light, runway turn off lights, wing scan lights, anti 2._____beacon, landing lights, 3. _____ lights, strobe lights, and logo light.

The external lighting controls are found on the forward section of the overhead panel. These lights can 4._____and off using the push buttons found here. Next, the strokes can be turned on using the 5._____. And the same to the taxi light. The runway turnoff lights can be turned on and off using the switches here. And if middle clicks, both switches can 6._____ simultaneously. Finally, for the left, right, and nose 7._____ lights, we switched them on individually, or they can be switched on and off simultaneously. Again 8._____ the middle click. The flood dick lighting controls are located in a number of positions, as 9._____. The master brightness control allows the number of flights and 10. _____ to be controlled using a single switch.

1.2 Aviation Speaking: practice the following dialogue and design your own.

Situation:

This is a dialogue between the pilot and the ground mechanic. There are obstacles when starting the engine or pushing back the aircraft. Practice it and have your own.

Note: PIL=Pilot, GND=Ground Mechanic

3 对话录音

Scene 1

GND: Cockpit, Ground.

PIL:　Go ahead.

GND: Standby for engine starting.

PIL:　Roger, Standby.

GND: BLOCK/BROKEN, please say again/repeatedly. I cannot catch your meaning/I do not know what you said. Please transfer the microphone.

PIL:　What's up?/ What's happened?

GND: A cargo trailer is behind the engine NO.2. Standby.

PIL:　Roger. Standby.

4 对话译文

Scene 2

G:　　Cockpit, Ground.

C:　　Go ahead.

G:　　Standby for pushing back. Traffic behind.

C:　　Roger, Standby.

G:　　Cockpit, Ground. Ready for pushing back.

C:　　Roger, pushing back.

Words & Expressions

1. block [blɒk] *v.*　　　　　　　阻塞
2. transfer [trænsˈfɜː(r)] *v.*　　　转让，转交

Section 2 Aviation Reading

Pre-reading questions:
1. Are different lights on the aircraft necessary?
2. What's the function of exterior lights?

5 课文朗读　　6 课文译文　　7 外部灯光系统 &
灯光子系统（微课）

Lighting System

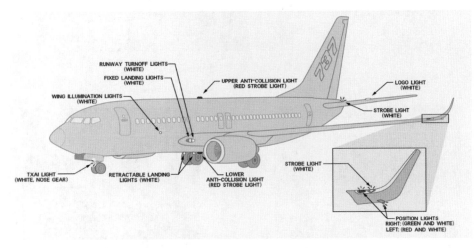

Fig.12-1　Schematic for B737-700 Lighting System

Introduction

Dynamic operational environments in aerospace industry demands high reliability and flexibility of the aircraft to overcome each operational challenge and to ensure safety of the passengers and crew. Effective usage of a wide combination of lights on the aircraft aids in addressing this challenge. Although the number of fatalities caused by improper usage of lights is comparatively lesser, lighting in an aircraft poses one of the major threats. Generally the overall lighting system in the aircraft can be differentiated into interior, exterior, service and emergency lighting system.

As competition among the existing aircraft operators has been increasing and the differentiating factor has been decreasing, aircraft operators are now focusing on enhancing the experience of the passengers during the travel. Interior lighting in the aircraft is one of the most important essentials of "the feel good" factor sensed by the passengers. This trend would change the dimension of overall interior lighting system in the near future.

On the other hand, exterior lighting system protect aircraft from possible mishaps such as plane crashes. For instance, anti-collision lights aid to spot every aircraft from any angle and also on the basis of light color the direction of the aircraft can be determined. Hence, the major functionality of exterior lighting system is to facilitate clear visibility for pilots and also to detect presence of aircraft and other obstacles in the sky. Exterior lighting system can be classified on the basis of application into aircraft visibility, pilot visibility and specific purpose lighting.

飞机维修专业英语——飞机系统（第二版）

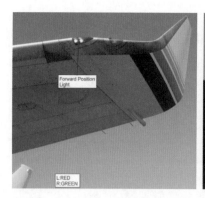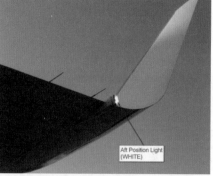

1&2. Forward &Aft Position light

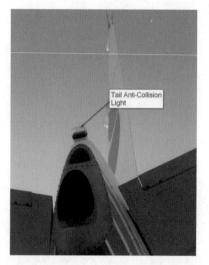

3. Tail Anti-collision Light   4. Taxi light

5.　Winglet Anti-collision Light   6. Lower Anti-collision Light

Fig.12-2　Exterior Lighting System

7. Runway Turnoff Light

8. Logo light

Fig.12-2　Exterior Lighting System（续图）

Emergency Lighting system is mostly used in signage in aircraft whereas specific purpose lighting finds their application mostly in cargo bays, wheel wells, equipment bays and fueling panels to aid the service personnel to carry out their tasks.

With the related to technology used in lighting, technological advancement has aided aircraft operators to choose from a wide range of available technologies such as LEDs, OLEDs, Incandescence, Electro/ Photo luminescent, fluorescent and strobe lights are the majorly used types of technologies. LEDs and OLEDs are one of the newest and most widely used types of lighting systems due to their light weight and high efficiency.

Subsystems of the Lighting System

The lighting system provides the necessary illumination for flight crew, passengers, service and cargo handling. It also provides lighting in emergency. The subsystems of the lighting system are as follows.

➢ Flight compartment lights.

➢ Passenger compartment lights.

➢ Cargo compartment lights.

➢ Service compartment lights.

➢ Exterior lights.

➢ Emergency lights.

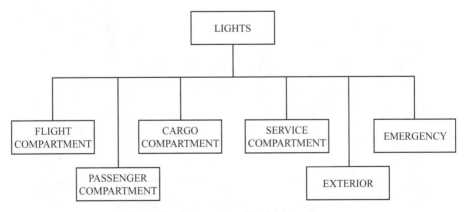

Fig.12-3　Subsystems of the Lighting System

Flight Compartment Lights

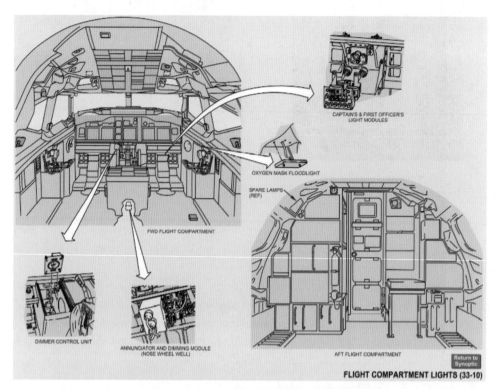

Fig.12-4　Different Lights in the Flight Compartment-1

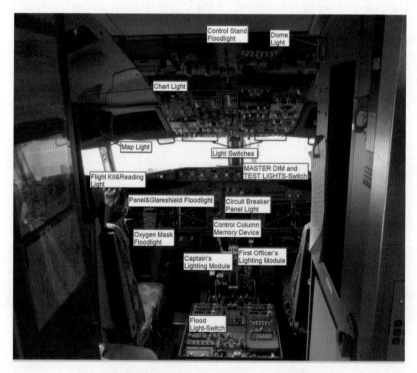

Fig.12-5　Different Lights in the Flight Compartment-2

These are the four types of flight compartment lights:

➢ Instrument and panel lights.

➢ Miscellaneous lights.

➢ Flight crew lights.

➢ Master dim and test.

The instrument and panel lights are for the flight compartment controls and panel indications. The miscellaneous lights supply general lighting to the flight compartment. These are the miscellaneous lights:

➢ Panel and control stand flood lights.

➢ Circuit breaker panel lights.

➢ Standby compass light.

➢ Dome lights.

The flight crew lights supply light for specific tasks. These are the flight crew lights:

➢ Reading lights.

➢ Map lights.

➢ Flight kit lights.

➢ Chart lights.

The master dim and test system controls the light for the system annunciator and indicator lights. Two white dome lights on the circuit breaker panels behind the crew provide general flight compartment area illumination. They are controlled by a switch on the aft overhead panel, identified as DOME LIGHT. This switch is provided with both bright and dim positions. A separate lamp is included in the left dome light that is part of the emergency lighting system.

The light-shield provides background lighting for the pilot main instrument panels. Each instrument and instrument panel has integral lighting. The control stand is illuminated from an overhead floodlight. Circuit breaker panels in the aft section of the flight compartment are illuminated by flood lights.

The master dim and test switch can do these functions:

➢ Testing the indicator and panel lights.

➢ Causing the indicator and panel lights to come on in the bright or dim mode.

Passenger Compartment Lights

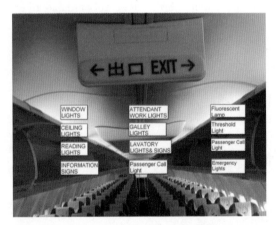

Fig.12-6　Different Lights in the Flight Compartment

The lights in the passenger compartment supply these functions:

➢ General light in the passenger cabin.

➢ Reading lights.

➢ Information lights for the passengers and attendants.

➢ Lights in the lavatories.

➢ Lights in the galleys.

The lighting system also lets the passengers call the attendants for aid. The passenger compartment lights have these subsystems:

➢ Window lights.

➢ Ceiling lights.

➢ Reading lights.

➢ Passenger signs.

➢ Lavatory lights and signs.

➢ Passenger and lavatory call lights.

➢ Attendant work lights.

➢ Galley lights.

➢ Entry lights.

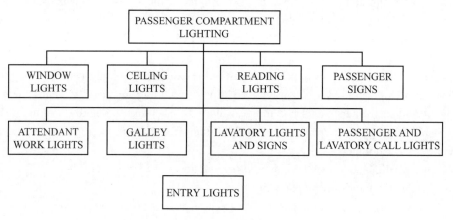

Fig.12-7   Passenger Compartment Lighting

You control these passenger cabin lights from the forward and aft attendant panels:

➢ Ceiling.

➢ Window.

➢ Entry.

➢ Work.

➢ Emergency (aft panel only).

The P 13 forward attendant panel is on the forward lavatory wall. The P 14 aft attendant panel is on the aft lavatory wall. The Attendant Control Panel (ACP) controls the passenger cabin lights. The ACP lighting display has these sections:

➢ Passenger seating area.

➢ Forward entry area.

➢ Aft entry area.

➢ General lighting layout.

The cargo and service compartment lights supply light to help maintenance personnel and ground crew. There are cargo compartments lights in the forward and aft cargo compartments. There are service lights in these areas:

➢ Forward equipment compartment.
➢ Electronic equipment compartment.
➢ Right air conditioning compartment.
➢ Left air conditioning compartment.
➢ Aft accessory compartment (section 48 interior).
➢ APU compartment.
➢ Tail cone compartment.

The Exterior Lights

The exterior lights supply light for airplane identification, direction, and to aid in the safe operation of the airplane.

These are the exterior lights on the airplane:

➢ Wing illumination light.
➢ Fixed landing lights.
➢ Anti-collision light(white).
➢ Anti-collision light(red).
➢ White, red and green position lights.
➢ Taxi and runway turnoff lights
➢ Logo lights.

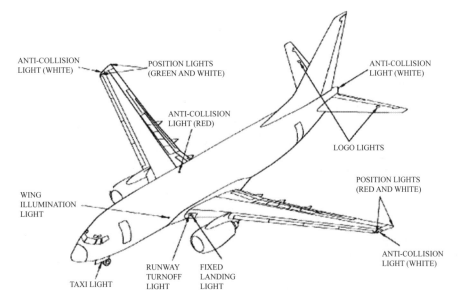

Fig.12-8   Exterior Lights-1

The emergency lighting system puts lights on areas inside and outside of the airplane. The emergency lights also show the exit paths.

The emergency lights operate at these moments:

➢ When emergency light system is on.
➢ When there is a loss of airplane DC power and the P 5 forward overhead panel emergency light switch is in the ARMED position.

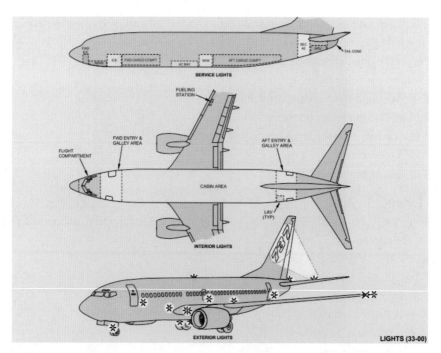

Fig.12-9   Exterior Lights-2

Control switches are at these locations:

➤ P 5 forward overhead panel in the flight compartment.

➤ P 14 aft attendant panel.

The emergency lighting system has these components:

➤ Exit signs.

➤ Aisle lights.

➤ Floor proximity lights.

➤ Slide lights.

➤ Power supplies.

Abbreviations & Acronyms

1.	Attendant (ATT)	服务人员
2.	Bright (BRT)	明亮的
3.	Electronic Equipment (EE)	电子设备
4.	Primary (PRIM)	首要的
5.	Typical (TYP)	典型的，特有的
6.	Variable (VAR)	变化的，可变的

Words & Expressions

1.	reliability [rɪˌlaɪəˈbɪlɪti] n.	可靠性
2.	flexibility [ˌfleksəˈbɪləti] n.	灵活性
3.	overcome [ˌəʊvəˈkʌm] v.	克服
4.	combination [ˌkɒmbɪˈneɪʃn] n.	结合
5.	address [əˈdres] v.	解决
6.	fatality [fəˈtæləti] n.	死亡

8 单词朗读

9 术语专讲：电门
（微课）

7. comparatively [kəm'pærətɪvli] *adv.*   相对地
8. lesser ['lesə(r)] *adj.*   较少的
9. pose [pəʊz] *v.*   引起
10. differentiate [ˌdɪfə'renʃieɪt] *v.*   区分
11. interior [ɪn'tɪəriə(r)] *n.*   内部
12. exterior [ɪk'stɪəriə(r)] *n.*   外部
13. decrease [dɪ'kriːs] *v.*   减少
14. enhance[ɪn'hɑːns] *v.*   提高
15. dimension [daɪ'menʃn] *n.*   规模
16. mishap ['mɪshæp] *n.*   小事故
17. spot [spɒt] *v.*   发现
18. facilitate [fə'sɪlɪteɪt] *v.*   促进
19. visibility [ˌvɪzə'bɪləti] *n.*   能见度
20. detect [dɪ'tekt] *v.*   发现
21. signage ['saɪnɪdʒ] *n.*   标志
22. incandescence [ˌɪnkæn'desns] *n.*   白炽
23. luminescent [ˌluːmi'nesnt] *adj.*   发光的
24. fluorescent [ˌflɔː'resnt] *n.*   荧光灯
25. strobe [strəʊb] *n.*   频闪灯
26. illumination [ɪˌluːmɪ'neɪʃn] *n.*   照明
27. handling ['hændlɪŋ] *v.*   操作，操纵
28. subsystem ['sʌbsɪstəm] *n.*   子系统
29. instrument ['ɪnstrəmənt] *n.*   仪表
30. miscellaneous [ˌmɪsə'leɪniəs] *adj.*   各种各样的
31. master ['mɑːstə(r)] *adj.*   主要的
32. dim [dɪm] *adj.*   暗的
33. indication [ˌɪndɪ'keɪʃn] *n.*   指示
34. flood light   探照灯，泛光灯
35. circuit breaker   断路器，跳开关
36. kit [kɪt] *n.*   装备
37. chart [tʃɑːt] *n.*   航图
38. annunciator [ə'nʌnsieɪtə(r)] *n.*   指示器，信号牌
39. switch [swɪtʃ] *n.*   开关，电门
40. lamp [læmp] *n.*   灯
41. dome [dəʊm] *n.*   圆顶，舱顶
42. attendant [ə'tendənt] *n.*   服务员
43. ceiling ['siːlɪŋ] *n.*   天花板
44. accessory [ək'sesəri] *n.*   附件
45. layout ['leɪaʊt] *n.*   布局
46. cone [kəʊn] *n.*   圆锥
47. anti-collision   防撞
48. turnoff light   转弯灯
49. aisle [aɪl] *n.*   过道
50. proximity [prɒk'sɪməti] *n.*   接近

Exercises

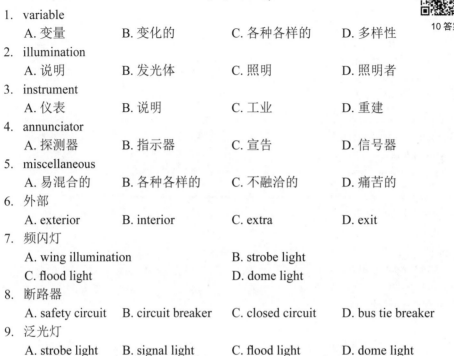

10 答案

2.1 Choose the best answer from A, B, C, D options.

1. variable
   A. 变量
   B. 变化的
   C. 各种各样的
   D. 多样性
2. illumination
   A. 说明
   B. 发光体
   C. 照明
   D. 照明者
3. instrument
   A. 仪表
   B. 说明
   C. 工业
   D. 重建
4. annunciator
   A. 探测器
   B. 指示器
   C. 宣告
   D. 信号器
5. miscellaneous
   A. 易混合的
   B. 各种各样的
   C. 不融洽的
   D. 痛苦的
6. 外部
   A. exterior
   B. interior
   C. extra
   D. exit
7. 频闪灯
   A. wing illumination
   B. strobe light
   C. flood light
   D. dome light
8. 断路器
   A. safety circuit
   B. circuit breaker
   C. closed circuit
   D. bus tie breaker
9. 泛光灯
   A. strobe light
   B. signal light
   C. flood light
   D. dome light
10. 转弯灯
    A. wing light
    B. anti-collision light
    C. logo light
    D. turnoff light

2.2 Find the best option to paraphrase the sentence.

1. Do not open the cargo door if the wind speed is dangerously high.
   A. Provided that the wind speed is dangerously high, do not open the cargo door.
   B. Do not open the cargo door unless there is dangerously high.
   C. Do not open the cargo door as if the wind speed is dangerously high.
   D. Do not open the cargo door even if the wind speed is very low.
2. A separate lamp is included in the left dome light that is part of the emergency lighting system.
   A. A inseparable lamp is included in the left dome light that is part of the emergency lighting system.
   B. A apparent lamp is included in the left dome light that is part of the emergency lighting system.
   C. There is a separate lamp in the left dome light that is part of the emergency lighting system.
   D. A alone lamp is included in the left dome light that is not part of the emergency lighting system.
3. The miscellaneous lights supply general lighting to the flight compartment.
   A. The miscellaneous lights apply general lighting to the cargo compartment.
   B. The various lights supply general lighting to the flight compartment.

C. The light supplies general lighting to the flight compartment.

D. The variety of lights supply emergency lighting to the flight compartment.

4. Although the number of fatalities caused by improper usage of lights is comparatively lesser, lighting in an aircraft poses one of the major threats.

A. Although a number of fatalities caused by improper usage of lights are comparatively more, but lighting in an aircraft poses one of the major threats.

B. If the number of fatalities caused by improper usage of lights is comparatively lesser, lighting in an aircraft poses one of the threats.

C. Despite the number of fatalities caused by improper usage of lights is comparatively more, lighting in an aircraft poses one of the major threats.

D. Though the number of fatalities caused by improper usage of lights is comparatively lesser, lighting in an aircraft poses one of the major threats.

5. Exterior lighting system can be classified on the basis of application into aircraft visibility, pilot visibility and specific purpose lighting.

A. Interior lighting system can be classified on the basis of application into aircraft visibility, pilot visibility and specific purpose lighting.

B. Exterior lighting system can be separated on the basis of application from aircraft visibility, pilot visibility and specific purpose lighting.

C. Exterior lighting system can be classic on the basis of application of aircraft visibility, pilot visibility and specific purpose lighting.

D. Exterior lighting system can be divided on the basis of application into aircraft visibility, pilot visibility and specific purpose lighting.

2.3　Change the following components specified in lower case and translate them into Chinese.

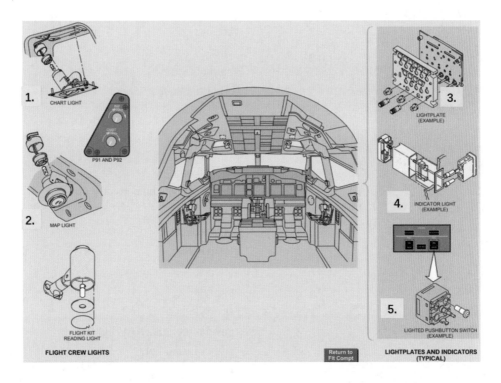

1.	
2.	
3.	
4.	
5.	

Section 3 Aviation Translation: translate the following sentences into Chinese

1. WARNING: DO NOT TOUCH THE SOCKETS OR THE METAL ENDS OF THE LAMP. WHEN ELECTRICAL POWER IS SUPPLIED TO THE LIGHT, ELECTRICAL SHOCK CAN OCCUR.

11 中国制造——
C919（文本）

2. WARNING: FOR THE LOWER ANTI-COLLISION LIGHT, MAKE SURE YOU SEAL EACH BARE ELECTRICAL CONNECTION NEAR THE LIGHT TO PREVENT AN EXPLOSION OF FUEL FUMES. AN EXPLOSION CAN CAUSE INJURIES TO PERSONS AND DAMAGES TO EQUIPMENT.

Section 4 Aviation Writing

Liu Shuai is an aircraft mechanic. He is inspecting the plane and finds some faults.

Please help him write down fault descriptions. Some related words, phrases & terms and key sentences are offered as follows.

Key words, phrases & terms:

1. 明的   bright
2. 暗的   dim
3. 烧坏   burn out
4. 灯亮   on / illuminate
5. 灯灭   off/ turn off
6. 不灭   does not go off
7. 闪亮   flash
8. 保持明亮   stay on bright
9. 不能恰当地点亮   does not come on correctly
10. 灯泡   bulb
11. 灯罩   lamp cover
12. 灯架   lamp holder
13. 灯座   lamp socket / lamp base
14. 尾灯   tail light
15. 闪光灯   strobe light
16. 航行灯   navigation light (NAV light)
17. 防撞灯   anti-collision light

18. 防撞（频闪）灯   anti-collision (strobe) light

19. 着陆灯   landing light

20. 信号灯   signal light

21. 下滑灯   approach light

22. 标志灯   logo light

23. 滑行灯   taxi light

24. 转弯灯   turnoff light

25. 机翼照明灯   wing illumination light

26. 位置灯（导航）   position (NAV) light

27. 信息灯   MSG ligh

28. 圆顶灯   dome light

29. 主警告灯   MASTER CAUTION light

Key sentences:

1. 航后检查发现 22 排左侧壁灯不亮。

   Do AF Check and find Row 22 Left window light does not work.

2. 右内着陆灯不亮。

   The R/H fixed landing light does not work.

3. L1 门 ENTRY 灯间歇性不工作，时亮时灭。

   The entry light of door L1 does not work intermittently.

4. 航后检查发现备用皮托管指示灯不亮。

   Do AF Check and find the ALT pitot indicator light does not work.

5. 前舱服务间的顶灯接触不好，时亮时暗。

   The ceiling light in FORWARD GALLEY does not work intermittently.

6. 7 排 C 座阅读灯坏。

   The reading light of Seat 7C does not work.

7. 航后检查发现前轮舱前壁灯不亮。

   Do AF Check and find the forward sidewall light in NLG wheel well does not work.

8. AF 检查发现左侧转弯灯不亮。

   The L/H runaway turnoff light is not illuminated in AF check.

9. 航后检查发现左右大翼照明灯均不亮。

   Neither the L/H nor R/H wing illumination lights are not illuminated in AF check.

10. 航后发现右侧翼根着陆灯不亮。

    The R/H inboard landing light does not illuminate in AF check.

11. 更换左侧根部的着陆灯。

    Replace L/H inboard landing light.

Task: Liu Shuai has three writing tasks. The first one is that the EXIT light in forward galley twin- kles all time on DIM position, the second is that the nose wheel well light does not come on during after flight check, and the last one is to replace right runway turnoff light. Please finish the writing tasks for him.

Warning Case

Eastern Airline 401- Lethal Bulb

12 警示案例——东航 401—致命灯泡

On December 29th, 1972, Eastern Airline 401 was a scheduled flight from John F. Kennedy International Airport to Miami International Airport. The aircraft operating the route, a Lockheed L1011-385-1 Tristar crashed into the Florida Everglades, resulting in 101 fatalities. The pilots and the flight engineer, two of the 10 flight attendants, and 96 out of the 163 passengers died. 75 passengers and crew survived.

Warning Tips: The crash occurred while the whole cockpit crew were preoccupied with a burnt-out landing gear indicator light. They failed to notice that autopilot had inadvertently disconnected and the aircraft started descending, and as a result, the plane smashed into the everglades.

Lesson 13　Navigation System

Learning Objectives

1. Knowledge objectives:

   A. To grasp the words, related terms and abbreviations about navigation system.

   B. To grasp the key sentences about navigation system.

   C. To know the main components about navigation system.

2. Competence objectives:

   A. To be able to read and understand frequently-used & complex sentence patterns, capitalized English materials and obtain key information about navigation system quickly.

   B. To be able to communicate with English speakers.

   C. To be able to fill in the job cards in English.

3. Quality objectives:

   A. To be able to self-study with the help of aviation dictionaries, the Internet and other resources.

   B. To develop the craftsman spirit of carefulness and responsibility.

Section 1　Aviation Listening & Speaking

1 听力录音　　2 答案及原文

1.1 Aviation listening: listen to the record and fill in the blanks.

What is that 1._____and 2._____you see on the runway just prior touch down? Let's first talk about the number.

We're going to have to 3._____some basic navigation first. As we look at our 4._____ earth, we have the geometric north and south pole in each hemisphere, marking the top and bottom of the planet. But there's no 5._____besides 6._____which could guide us towards these poles.

But the planet has its own magnetic field surrounding the earth, creating the magnetic north and south poles which are 7._____off of the geometric poles.

Now you've all seen a 8._____, and you know that the needle always points towards magnetic north.

Now let's look at this runway here, you can see the numbers 08, let's 9._____you would draw a line all the way from the magnetic north pole down until it crosses the runway 10._____. You would see that the two lines create an opening angle of 080 degrees. Meaning the runway centerline is 080 degrees relative to the magnetic north pole. So why is there only a 08 on the runway? There are two reasons for that. Let's say the magnetic heading of the runway would be 084. Then it would be rounded off to the nearest tenth, so in this case, 080, leaving out the last zero, you have runway "zero eight". But if you would name the runway "zero eight zero", it could lead to misinterpretation as given headings are spelled in the same way, like fly heading 120 for example.

Now as you sit in your plan on the runway, in this example runway 05 in Madeira you know your plane nose is pointing 050 degrees, so heading northeast, and your airplane tail is pointing + 180 degrees into the opposite direction, so the reciprocal runway should be? Yes, runway "two three".

1.2 Aviation Speaking: practice the following dialogue and design your own.

Situation:

This is a dialogue between the pilot and the air traffic controller. After starting, the crew call the ground that there was a fault in the cockpit. Practice it and have your own.

3 对话录音

Note: PIL=Pilot, CTL= Air Traffic Controller

Scene 1

PIL:　　Hong Kong Approach CSN 303, we have been struck by lightning, rear fuselage damaged, right elevator lost, request priority landing and radar vector to final approach.

CTL:　　CSN 303 you are number one, radar will assist you until landing.

Scene 2

PIL A:　Exterior check completed.

PIL B:　Everything in order out there?

PIL A:　All good, Captain.

PIL B:　Metro-jet 9248. Request permission to taxi.

CTL:　　Metro-jet 9248, runway 04R.

PIL:　　Metro-jet 9248 requests takeoff clearance, runway 04R.

CTL:　　Cleared for take-off runway 04R. The airbus seems to be dropping.

Tower:　Metro-jet 9248，are you experiencing any difficulties? Do you read me, Metro-jet 9248?

4 对话译文

Words & Expressions

1. CSN　　　　　　　　　　　　　　　中国南方航空
2. priority [praɪˈɒrəti] *n.*　　　　　　优先
3. exterior [ɪkˈstɪəriə(r)] *adj.*　　　　外部的
4. permission [pəˈmɪʃn] *n.*　　　　　准许
5. clearance [ˈklɪərəns] *n.*　　　　　许可

Section 2　Aviation Reading

Pre-reading questions:

1. What is navigation system?
2. Why does an airplane need a navigation system?

5 课文朗读

6 课文译文

7 飞机上的导航系统（微课）

Navigation System

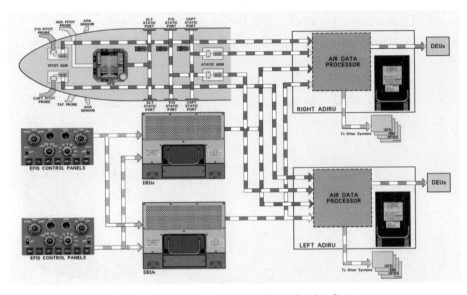

Fig.13-1　Schematic for B737-700 Navigation System

Introduction

Navigation is a field of study that focuses on the process of monitoring and controlling the movement of a craft or vehicle from one place to another. The field of navigation includes four general categories: land navigation, marine navigation, aeronautic navigation and space navigation. It is also the term used for the specialized knowledge used by navigators to perform navigation tasks. All navigational techniques involve locating the navigator's position compared to known locations or patterns.

The navigation system on the airplane is used to determine and display the attitude, altitude, and position of the airplane. The navigation system have the following categories:

➢ Those systems which sense and display flight environmental data.

➢ Those systems which determine airplane attitude and direction.

➢ Those systems which provide landing and taxiing aids.

➢ Those systems which are self-contained and independent of ground-based equipment.

➢ Those systems which are dependent upon and operate in conjunction with ground-based equipment.

➢ Those systems which compute airplane position.

The various navigation units are located in four main equipment centers throughout the airplane and are remotely controlled and displayed from the flight compartment. They are forward equipment center, main equipment center, forward center fuselage equipment center and aft center fuselage equipment center.

Receiving and/or transmitting antennas required by navigation systems are located externally on the airplane as shown in Fig.13-2.

Navigation system information is displayed on the primary flight displays (PFDs), Navigation Displays (NDs), and the captain's Radio Magnetic Indicator (RMI). The PFDs and NDs are described in the integrated display system. The RMI is described separately under Inertial Reference System, VOR Navigation System and ADF System.

Navigation system for aircraft has evolved with the nature and role of the aircraft itself. Whatever the requirement is, all navigation systems are concerned with several key factors:

➤ Accuracy: conformance between calculated and actual position of the aircraft.

➤ Integrity: ability of a system to provide timely warnings of system degradation.

➤ Availability: ability of a system to provide the required function and performance.

➤ Continuity: probability that the system will be available to the user.

➤ Coverage: geographic area where each of the above are satisfied at the same time.

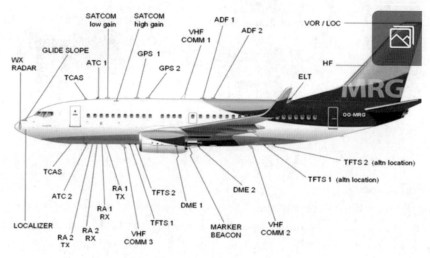

Fig.13-2　Receiving and/or Transmitting Antennas

Navigation Systems or Instruments on the Airplane

In the early days of airplane operation, navigation instruments either did not exist or, at most, consist of a magnetic compass and an airspeed indicator. When flying by visual reference, the early pilot would usually navigate from one landmark to another, following roads and railroads or rivers and valleys. Flights were made at comparatively low altitudes providing a view of the ground that was usually good enough for the pilot to clearly identify objects there. Under the flying conditions that existed when the airplane was considered a novelty, complex navigation instruments and systems were not in great demand. As the use of airplanes increased and flights were made at higher altitudes, above the clouds and at night, it became necessary to develop reliable navigation techniques along with instruments indicating attitude, heading, airspeed, and drift so that the pilot could determine the airplane.

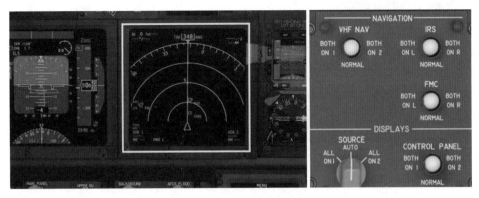

Fig.13-3   Navigation Display

Next some main systems or components in respect to navigation are illustrated.

Air Data Inertial Reference System (ADIRS)

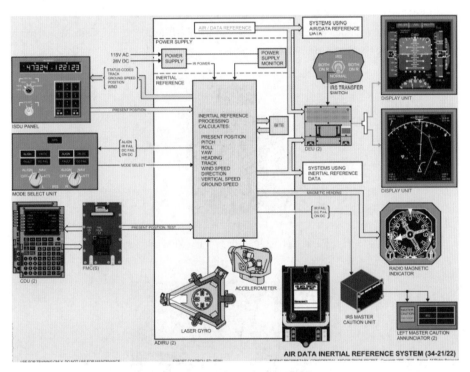

Fig.13-4   Schematic of ADIRS

The ADIRS supplies the data of altitude, attitude, airspeed, temperature, heading and present position to the aircrew and to the airplane systems. It has the following components:

➤ Air Data Modules (ADMs) (4).
➤ Total Air Temperature (TAT) probe.
➤ Angle of Attack (AOA) sensors (2).
➤ Inertial System Display Unit (ISDU).
➤ Mode Select Unit (MSU).
➤ Air Data Inertial Reference Unit (ADIRU) (2).
➤ IRS master caution unit.

> Alternate VMO/MMO Switch.

It uses these inputs to calculate air data:

> Pitot pressure.
> Static pressure.
> Total air temperature.
> Angle of attack.
> Common Display System (CDS) barometric correction.
> IR data.

There are ADIRS components of Inertial System Display Unit (ISDU), Mode Select Unit (MSU), IRS master caution annunciator in the flight compartment.

VHF Omni-directional Video Range (VOR)

The VOR systems receive, decode and process bearing information from the transmitted VOR signal. VOR means that it uses frequencies in the VHF band and has both omni and directional transmitted signals. You can compare the VOR principle with a lighthouse. It has a rotating beam and a flashlight which you can see in all directions. It flashes when the rotating beam points to magnetic north. When you measure the time between the flashlight and the visibility of the rotating beam you can identify the direction to the lighthouse.

Fig.13-5　VHF Omni-directional Video Range

Fig.13-6　Marker Beacon Antenna

Instrument Landing System (ILS)

The Multi-mode Receiver (MMR) contains the Instrument Landing System (ILS), the Global Navigation Satellite System (GNSS), the GNSS Landing System (GLS) and the global positioning system functions. The GLS provides lateral and vertical position data in an ILS look-alike manner necessary to put the airplane on the runway for approach. The system uses signals from satellites and Ground Based Augmentation System (GBAS) ground station. The GNSS receives signals through the GPS antenna and the GBAS ground station to provide aircraft position. The GBAS receives signals through the VOR antenna to provide GNSS differential corrections and Final Approach Segment (FAS).

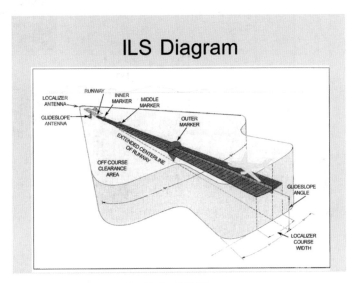

Fig.13-7   ILS Diagram

The ILS provides lateral and vertical position data necessary to put the airplane on the runway for approach. The system uses signals from a glideslope ground station and a localizer ground station. The glideslope ground station transmits signals to give the airplane a descent path to the touchdown point on the runway. The localizer ground station transmits signals to give the airplane lateral guidance to the runway centerline. The ILS has two Multi-mode Receivers (MMRs) that contain the ILS function. The ILS function in the MMRs receives inputs through these components:

➢ The receivers geting manual tune inputs from the Navigation (NAV)control panels.
➢ The VOR/LOC antenna and the localizer antenna sending localizer signals through the localizer antenna switches to the MMRs.
➢ The localizer antenna switches selecting the VOR/LOC antenna or the localizer antenna as the source of Radio Frequency (RF) input to the MMR.
➢ The glideslope antenna sending glideslope signals to the MMRs.

The marker beacon system supplies visual and aural indications when the airplane flies over airport runway marker beacon transmitters. The marker beacon system has an antenna and a VOR/ Marker Beacon (VOR/MB) receiver. The antenna sends RF signals to the VOR/ MB receiver 1. The Proximity Switch Electronics Unit (PSEU) supplies an air/ground discrete signal to the VOR/ MB receiver 1. The VOR/MB receiver uses this signal for the functions of preventing a test in the air and counting flight legs.

The marker beacon function only operates in the VOR/MB receiver 1 position. The antenna receives the marker beacon signals. The signals go to the VOR/MB receiver 1 and receiver 1 supplies this data:

➢ Marker beacon audio to the Remote Electronics Unit (REU).
➢ Marker beacon data to the Common Display System Display Electronics Unit (CDSDEU).
➢ Marker beacon data to the Flight Data Acquisition Unit (FDAU).

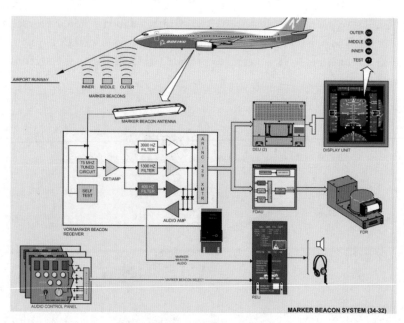

Fig.13-8　Marker Beacon System

Localizer(LOC)

The LOC ground station transmits 90 and 150 Hz signals for lateral guidance. These two signals are radiated to produce two directional lobes side by side along the runway centerline.

The left lobe is modulated with a 90 Hz signal and the right lobe with 150 Hz signal. The LOC ground station uses one of 40 channels in the 108.10 to 111.95 Megahertz frequency range but only frequencies with odd tenths.

The transmitter is coupled to an array of directional antennas that radiate the two lobes along the runway centerline. The transmitter shed and the antennas are located at the end of the runway. An aircraft flying down the centerline would receive a signal with equal levels of both modulations. This is shown by the centered deviation pointer on the indicator, here the NAV display.

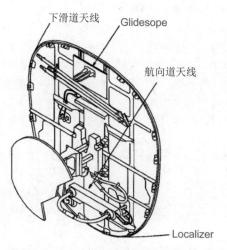

Fig.13-9　Localizer & Glideslope Antenna

If the Aircraft position is left of the centerline, the 90 Hz signal predominates and the LOC

deviation pointer deflects to the right, indicating that the runway centerline is to the right. One dot on the indicator normally shows a one degree offset on the LOC.

The Traffic Alert and Collision Avoidance System(TCAS) is designed to protect the airspace around a TCAS equipped airplane. Its function is to determine the range, altitude, bearing and closure rate of other aircraft which are equipped with an ATC transponder. TCAS monitors the trajectory of other aircraft to determine if there is any danger of a collision and provides the pilots with aural and visual advisories for a vertical avoidance maneuver.

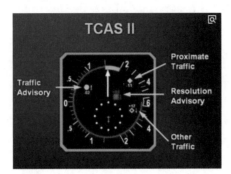

Fig.13-10　TTCAS

An Automatic Direction Finder (ADF) is an aircraft radio navigation system which senses and indicates the direction to a low/medium frequency Non-Directional Radio Beacon (NDRB) ground transmitter operating within range 190-1,750 kHz, i.e. low and medium frequency bands.

By FAA definition, ADF is: "electronic navigation equipment that operates in the low/medium frequency bands (200 nm). Used in conjunction with the ground-based Non-Directional Beacon (NDB), the instrument displays the number of degrees clockwise from the nose of the aircraft to the station being received." It is used to find the direction of not only a radio source (i.e: a radio tower), but also to use the bearings (locations) of two radio sources and the headings (directions of travel) of their signals in order to pinpoint the location of the aircraft receiving them. An example of one type of ADF indicator is the Radio Magnetic Indicator (RMI).

Fig.13-11　Automatic Direction Finder (ADF)

For the beginning pilot, particularly those flying in aircraft without more advanced navigation systems, an ADF system represents a component for in-flight navigation over long distances. For an aircraft maintenance engineer, particularly one specializing in an aircraft's electrical systems, it is a system that leaves no room for missteps in its upkeep and repair, where even the smallest of mistakes on their part could result in a margin of error literally miles wide.

Radio Magnetic Indicator (RMI)

The RMI is a standby instrument. Its purpose is to show relative bearing to VOR and ADF stations. It also shows airplane magnetic heading. The RMI receives digital heading data from the Air Data Inertial Reference Units (ADIRUs). The heading data sets the compass card to the magnetic heading of the airplane.

The RMI uses the heading data from the left ADIRU with the IRS transfer switch in the normal or both on left position. Move the switch to the both on right position to use heading data from the right ADIRU. The RMI receives digital bearing data from the VOR and ADF receivers. The RMI uses the bearing data to set the position of the bearing pointers.

An example of one type of ADF indicator is the RMI. A RMI consists of a magnetic compass backing with two indication needles: one, generally double-bared or thicker, and another thinner needle, both of which are tuned via the ADF to home into the signals being transmitted from two separate radio stations as the aircraft approaches them from the air. The small orange symbol at the center along with the orange arrow aligned with it and set into the compass degrees indicate the current orientation of the aircraft, in relation to the signals being received.

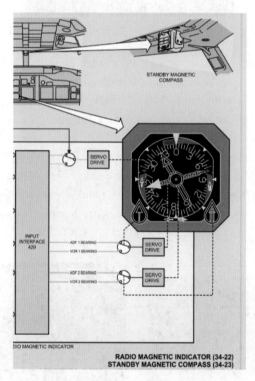

Fig.13-12　RMI

Abbreviations & Acronyms

1. Air Data Inertial Reference Unit (ADIRU)　大气数据惯性基准组件
2. Air Data Module (ADM)　　　　　　　　大气数据组件
3. Air Data Inertial Reference System(ADIRS) 大气数据惯性基准系统
4. Aeronautical Radio Incorporated (ARINC)　航空无线电公司
5. Amplifier (AMP)　　　　　　　　　　　放大器
6. Automatic Direction Finder (ADF)　　　　姿态测向仪

7. Common Display System (CDS)　　　　公共显示系统
8. Display Electronics Unit (DEU)　　　　显示电子组件
9. Distance Measuring Equipment (DME)　　测距设备
10. Flight Control Computer (FCC)　　　　飞行控制计算机
11. Flight Data Acquisition Unit (FDAU)　　飞行数据采集组件
12. Flight Management Computer (FMC)　　飞行管理计算机
13. Instrument Landing System (ILS)　　　仪表着陆系统
14. Inertial Reference System (IRS)　　　惯性基准系统
15. Marker Beacon (MB)　　　　　　　　指点信标
16. Navigation Display (ND)　　　　　　　导航显示
17. Primary Flight Display (PFD)　　　　　主飞行显示
18. Receiver (RCVR)　　　　　　　　　　接收机
19. Radio Magnetic Indicator (RMI)　　　无线电磁指示器
20. Remote Electronics Unit (REU)　　　　远程电子组件
21. VHF Omni-directional Vidio Range (VOR) 甚高频全向信标

Words & Expressions

1. navigation[ˌnævɪˈɡeɪʃn] *n.*　　　　导航
2. marine [məˈriːn] *adj.*　　　　　　航海的
3. aeronautic [ˌeərəˈnɔːtɪk] *adj.*　　航空的
4. attitude [ˈætɪtjuːd] *n.*　　　　　姿态
5. altitude [ˈæltɪtjuːd] *n.*　　　　海拔，高度
6. conjunction [kənˈdʒʌŋkʃn] *n.*　　结合
7. magnetic [mæɡˈnetɪk] *adj.*　　　磁性的
8. integrated [ˈɪntɪɡreɪtɪd] *adj.*　　集成的
9. inertial[ɪˈnɜːʃl] *adj.*　　　　　　惯性的
10. evolve [iˈvɒlv] *v.*　　　　　　　发展
11. accuracy [ˈækjərəsi] *n.*　　　　准确性
12. conformance [kənˈfɔːməns] *n.*　一致性
13. integrity [ɪnˈteɡrəti] *n.*　　　　完整
14. degradation [ˌdeɡrəˈdeɪʃn] *n.*　退化
15. availability [əˌveɪləˈbɪləti] *n.*　有效性
16. continuity [ˌkɒntɪˈnjuːəti] *n.*　　连续性
17. coverage [ˈkʌvərɪdʒ] *n.*　　　　覆盖范围
18. geographic [ˌdʒiːəˈɡræfɪk] *adj.*　地理的
19. landmark [ˈlændmɑːk] *n.*　　　　地标
20. valley [ˈvæli] *n.*　　　　　　　　山谷
21. comparatively [kəmˈpærətɪvli] *adj.*　对比地
22. reliable [rɪˈlaɪəbl] *adj.*　　　　　可靠的
23. drift [drɪft] *n.*　　　　　　　　偏航，偏移
24. pitot [ˈpɪtəʊ] *n.*　　　　　　　空速管，皮托管
25. static [ˈstætɪk] *adj.*　　　　　静态的
26. barometric [ˌbærəˈmetrɪk] *adj.*　大气压力
27. annunciator [əˈnʌnsieɪtə(r)] *n.*　信号器
28. omni [ˈɒmnɪ] *n.*　　　　　　　全方位

8 单词朗读

9 术语专讲：雷达
波瓣（微课）

29. beam [biːm] *n.*   光束

30. visibility [ˌvɪzə'bɪləti] *n.*   能见度

31. lateral ['lætərəl] *n.*   横向的

32. glideslope [glaid'sləup] *n.*   下滑道

33. localizer ['ləʊkəlaɪzə] *n.*   定位器

34. descent [dɪ'sent] *n.*   下降

35. touchdown ['tʌtʃdaʊn] *n.*   着陆

36. beacon ['biːkən] *n.*   信标

37. discrete [dɪ'skriːt] *adj.*   离散的

38. acquisition [ˌækwɪ'zɪʃn] *n.*   获得

39. array [ə'reɪ] *n. & v.*   排列

40. radiate ['reɪdieɪt] *n.*   有射线的

41. lobe [ləʊb] *n.*   波瓣

42. deviation [ˌdiːvi'eɪʃn] *n.*   偏差

43. predominate [prɪ'dɒmɪneɪt] *v.*   占支配地位

44. deflect [dɪ'flekt] *v.*   使偏斜，使转向

45. offset ['ɒfset] *v.*   抵消，补偿

46. closure rate   终止率

47. transponder [træn'spɒndə(r)] *n.*   应答器

48. trajectory [trə'dʒektəri] *n.*   轨迹

49. misstep [ˌmɪs'step] *n.*   错误

50. align [ə'laɪn] *v.*   对齐

Exercises

10答案

2.1 Choose the best answer from A, B, C, D options.

1. IRS
   - A. 惯性基准系统   B. 大气数据惯性基准系统
   - C. 主飞行显示器   D. 导航显示器

2. ADF
   - A. 空速指示器   B. 自动定向机   C. 磁罗盘   D. 主警告系统

3. VOR
   - A. 迎角   B. 公共显示组件   C. 总温探头   D. 甚高频全向信标

4. AOA
   - A. 迎角   B. 自动定向机   C. 主飞行显示器   D. 导航显示器

5. CDS
   - A. 自动定向机   B. 主警告系统   C. 公共显示系统   D. 磁罗盘

6. 惯性的
   - A. magnetic   B. inertial   C. airspeed   D. compass

7. 罗盘
   - A. static   B. magnetic   C. campus   D. compass

8. 静压
   - A. static pressure   B. kinetic pressure
   - C. differential pressure   D. pitot pressure

9. 全压

    A. kinetic pressure   B. pitot pressure

    C. differential pressure   D. static pressure

10. 姿态

    A. altitude   B. latitude   C. attitude   D. attendant

2.2 Find the best option to paraphrase the sentence.

1. It is also the term used for the specialized knowledge used by navigators to perform navigation tasks.

    A. It is also the term used for the universal knowledge used by navigators to perform navigation tasks.

    B. It is also the term used for the specialized knowledge used by navigators to execute navigation tasks.

    C. It is also the term used for the special knowledge used by navigators to perform navigation tasks.

    D. It isn't the term used for the specialized knowledge used by navigators to perfect navigation tasks.

2. Next some main systems or components in respect to navigation are illustrated.

    A. Next some main systems or components with regard to navigation are illustrated.

    B. Next some main systems or compositions in respect to navigation are illustrated.

    C. Next some main systems or components related to flight controls are illustrated.

    D. Next some main systems or components irrelated to navigator are illustrated.

3. It is compulsory to do the procedure in this service bulletin.

    A. It is optional to do the procedure in this service bulletin.

    B. It is unnecessary to do the procedure in this service bulletin.

    C. It is mandatory to do the procedure in this service bulletin.

    D. It is selective to do the procedure in this service bulletin.

4. Do not open the cargo door if the wind speed is dangerously high.

    A. Do not open the entry door if the wind speed is dangerously high.

    B. Do not open the cargo door unless there is dangerous wind.

    C. Do not open the cargo door when the wind speed is dangerously high.

    D. Do not open the cargo door even if there is dangerous wind.

5. This document provides supplemental information to the applicable airplane maintenance manual.

    A. This document provides optional information to the suitable airplane maintenance manual.

    B. This document supplies supplemental information to the applicable airplane maintenance manual.

    C. This document provides extra information to the accessible airplane maintenance manual.

    D. This document provides excessive information with the applicable airplane maintenance manual.

2.3 Change the following components specified in lower case and translate them into Chinese.

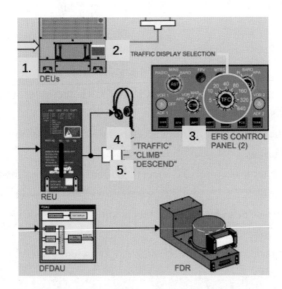

1.	
2.	
3.	
4.	
5.	

Section 3 Aviation Translation: translate the following sentences into Chinese

1. CAUTION: MAKE SURE YOU DO NOT SUPPLY ELECTRICAL POWER TO THE PITOT PROBE HEATER. THIS CAN CAUSE DAMAGES TO THE PITOT PROBE.

11 情注蓝天 航修
报国—李天（文本）

2. WARNING: DO NOT OPERATE THE WEATHER RADAR WHILE FUEL IS ADDED OR REMOVED FROM THE AIRPLANE. DO NOT TRANSMIT RF ENERGY WHILE FUEL IS ADDED OR REMOVED IN AN AREA 300 FEET OR LESS IN FRONT OF THE ANTENNA. THIS CAN CAUSE AN EXPLOSION.

Section 4 Aviation Writing

Liu Shuai is an aircraft mechanic. He is inspecting the plane and finds some faults.

Please help him write down fault descriptions. Some related words, phrases & terms and key sentences are offered as follows.

Key words, phrases & terms:

1. 导航参数　　　　　　　　　navigational parameter
2. 航向　　　　　　　　　　　heading
3. 磁航向　　　　　　　　　　magnetic heading
4. 真航向　　　　　　　　　　true heading
5. 航迹角　　　　　　　　　　track angle
6. 偏流角　　　　　　　　　　drift angle
7. 航道角　　　　　　　　　　flight path angle
8. 地速　　　　　　　　　　　Ground Speed(GS)
9. 全压　　　　　　　　　　　pitot pressure
10. 静压　　　　　　　　　　　static pressure
11. 动压　　　　　　　　　　　impact pressure
12. 静压口　　　　　　　　　　static ports
13. 皮托管　　　　　　　　　　pitot probe172
14. 校正气压高度　　　　　　　corrected carometric altitude
15. 大气静温　　　　　　　　　Static Air Temperature(SAT)
16. 大气全温　　　　　　　　　Total Air Temperature(TAT)
17. 惯性高度　　　　　　　　　inertial altitude
18. 气象雷达故障　　　　　　　Weather Radar (WXR) FAIL
19. 地形　　　　　　　　　　　terrain
20. 气象雷达系统　　　　　　　WXR System
21. 气象雷达天线　　　　　　　WXR antenna
22. 气象雷达收发机支架　　　　WXR R/T mount
23. 气象雷达 / 地形继电器　　　WXR/TERR Relays
24. 指点信标系统　　　　　　　marker beacon system
25. 指点信标天线　　　　　　　marker beacon antenna
26. 指点信标音频　　　　　　　marker beacon audio

Key sentences:

1. 静压系统渗漏测试。

   Static system low-range leak test.

2. 动压系统渗漏测试。

   Pitot system leak test.

3. 航后根据 AMM 34-11-01 更换机长皮托管，根据 AMM 34-11-00 进行渗漏测试，结果符合手册要求。

   AF: Refer to AMM 34-11-01 to replace the capt pitot probe, and refer to AMM 34-11-00 to do the left pitot system leak test, the results satisfy to AMM requirements.

4. 机组反映备用高度表与 CDS 误差较大。

   The flight crew report the altitude differences between the CDS and the standby altitude indicator are large/obvious.

5. 航后根据 AMM 34-13-01 更换备用高度空速表，根据 AMM 34-13-00 进行打压测试，结果符合手册要求。

   AF: Refer to AMM 34-13-01 to replace the standby altitude/airspeed indicator, and refer to AMM 34-13-00 to do the system test, the results satisfy to AMM requirements.

6. 经过导航台时屏幕无显示。

Indication does not show when passing a marker.

7. RA 收发机前面板 "LRU STATUS" 灯亮。

RA Receiver/Transmitter "LRU STATUS" light on.

8. 气象雷达范围不一致。

WXR range disagree.

9. 信标音频无声音或较弱。

Audio tone does not occur or is weak.

Task: Liu Shuai has three writing tasks. The first one is that there is no indication when passing a marker, the second is that there is weak or no audio tone, and the last one is that weather range does not agree. Please finish the writing tasks for him.

Warning Case

Korean Air Cargo 8509-Bad Attitude

12 警示案例——大韩航空
货运 8509——刚愎自用

On December 22nd , 1999, 55 seconds after takeoff, Korean air cargo 8509 crashed into the ground. The plane finally crashed, with no ground casualties, but all four crew members were killed, including a Korean aviation aircraft maintenance technician.

Warning Tips: After learning that the attitude indicator failed, the maintenance personnel did not replace it according to the instructions of AMM, but just reconnected the wire and reset it. Finally, the pilot mistakenly believed the display, causing the plane to tilt seriously and crash. The crew in the accident lacked communication. Before the accident, the flight engineer had pointed out that the aircraft tilted seriously and the alarm sounded loudly, but the co-pilot chose to ignore the alarm and failed to correct the captain's mistake in time, which eventually led to the accident.

Lesson 14 Warning System

Learning Objectives

1. Knowledge objectives:

   A. To grasp the words, related terms and abbreviations about warning system on an aircraft.

   B. To grasp the key sentences.

   C. To know the main components of warning system.

2. Competence objectives:

   A. To be able to read and understand frequently-used & complex sentence patterns, capitalized English materials and obtain key information about warning systems quickly.

   B. To be able to communicate with English speakers about the topic freely.

   C. To be able to fill in the job cards in English.

3. Quality objectives:

   A. To be able to self-study with the help of aviation dictionaries, the Internet and other resources.

   B. To develop the craftsman spirit of carefulness and responsibility.

Section 1 Aviation Listening & Speaking

1.1 Aviation Listening: listen to the record and fill in the blanks.

1 听力录音   2 答案及原文

A Ground Proximity Warning System (GPWS) is a 1.＿＿＿＿designed to alert pilots if their aircraft is in 2.＿＿＿＿danger of flying into the ground or an obstacle. The United States Federal Aviation Administration (FAA) defines GPWS as a type of 3.＿＿＿＿Awareness Warning System (TAWS). More advanced systems, introduced in 1996, are known as Enhanced Ground 4.＿＿＿＿ Warning System (EGPWS), although sometimes called terrain awareness warning system.

In the late 1960s, a series of 5.＿＿＿＿Flight Into Terrain (CFIT) accidents took the lives of hundreds of people. A CFIT accident is one where a properly functioning airplane under the control of a fully 6.＿＿＿＿and certified crew is flown into terrain, water or obstacles with no apparent awareness on the part of the crew.

Beginning in the early 1970s, a number of 7.＿＿＿＿examined the occurrence of CFIT accidents. Findings from these studies 8._____ that many such accidents could have been avoided if a warning device called GPWS had been used. As a result of these studies and recommendations from the U.S. National 9._____ Safety Board (NTSB), in 1974 the FAA required all large turbine and turbojet airplanes to install TSO-approved GPWS equipment. The ICAO recommended the installation of GPWS in 1979. In March 2000, the U.S. FAA amended operating rules to require that all U.S. registered turbine-powered airplanes with six or more passenger seats (exclusive of pilot and 10._____ seating) be equipped with a FAA-approved TAWS. The mandate affects aircraft manufactured after March 29, 2002.

1.2 Aviation Speaking: practice the following dialogue and design your own.

Situation:

This is a dialogue between the pilot and the air traffic controller. After starting, the crew report that the warning light is on. Practice it and have your own.

3 对话录音

Note: PIL=Pilot, CTL= Air Traffic Controller.

Scene 1

PIL: CSN 303 warning light is on, reason unknown, request first priority landing.

CTL: CSN 303, you are number one, turn left heading 270 and descend to 4,500 feet.

4 对话译文

Scene 2

CTL: Trans Asian 222, heading 240 approved. You are next in line, say your attention.

PIL: Trans Asian 222 request runway 20 VOR.

CTL: Trans Asian 222 Roger. Fly heading 02, 360 radar vector. VOR approach.

PIL: Ladies and gentlemen, we are about to begin our approach.

CTL: TERRAIN. TERRAIN. PULL UP. Go around. Trans Asian 222, do you copy?

Words & Expressions

1. descend [dɪˈsend] v.   下降
2. terrain [təˈreɪn] n.   地形

Section 2 Aviation Reading

Pre-reading questions:

1. How would the pilot know if a plane catches fire?
2. What indications does the warning system provide?

5 课文朗读   6 课文译文

Warning System

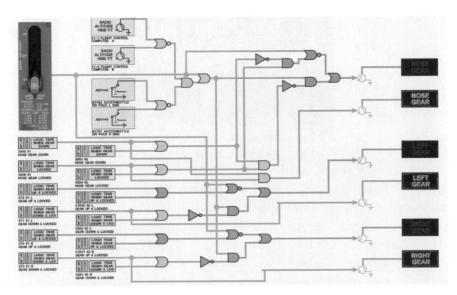

Fig.14-1 Schematic for B737-700 Warning System

The warning system provides the flight crew visual and aural indication of abnormal airplane systems conditions or that the airplane is out-of-configuration for specific operating modes, such as landing and takeoff. Input signals are supplied from airplane sensors, mechanical and avionic systems and the pilots. Outputs are aural tones from dedicated aural warning speaker in the flight deck and visual displays on flight compartment panels.

7 飞机上的警告
系统（微课）

Visual Warning Indication

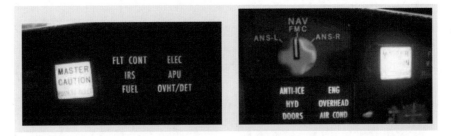

Fig.14-2 Master Caution & System Annunciator Lights(Left & Right)

Visual indication includes master warning lights, warning configuration light, and ICAS warning messages. The master warning lights located at both ends of glareshield illuminate for level A warnings associated with the following six conditions:

➢ Fire.
➢ Overspeed.
➢ Autopilot.
➢ Takeoff and landing configuration.
➢ Cabin altitude.
➢ Ground proximity warning mode 1 and 2.

Fig.14-3   GND PROX Warning

Each light contains a manual reset circuit for extinguishing both lights. Pressing either light switch provides a ground signal to master warning module to reset both lights. The CONFIG light located on the captain's instrument panel is a dual lamp red indicator which continuously illuminates when the configuration warning module sense improper takeoff or landing configuration. The light extinguishes when condition is cleared.

The master caution system was developed for the 737 to ease pilot workload as it was the first Boeing airliner to be produced without a flight engineer. In simple terms it is an attention getter that also directs the pilot toward the problem area concerned. The system annunciators (shown above) are arranged such that the cautions are in the same orientation as the overhead panel e.g. fuel bottom left, doors bottom of third column, etc.

Before engine start, use individual system lights to verify the system status. If an individual system light indicates an improper condition.

On the ground, the master caution system will also tell you if the condition is dispatchable. The FCOM gives the following guidance on master caution illuminations on the ground:
   ➢ Check the Dispatch Deviations Procedures Guide (DDPG) or the operator equivalent to decide if the condition has a dispatch effect.
   ➢ Decide if maintenance is needed.
If, during or after engine start, a red warning or amber caution light illuminates:
   ➢ Do the respective Non-Normal Checklist (NNC).
On the ground, check the DDPG or the operator equivalent.
   ➢ If, during recall, an amber caution illuminates and then extinguishes after a master caution reset:
   ➢ Check the DDPG or the operator equivalent.
   ➢ The respective non-normal checklist is not needed.
Pressing the system annunciator will show any previously cancelled or single channel cautions.

If a single channel caution is encountered, the QRH drill should not be actioned.

Master caution lights and the system annunciator are powered from the battery bus and will illuminate when an amber caution light illuminates. Exceptions to this include a single centre fuel tank LOW PRESSURE light (requires both), REVERSER lights (requires 12 seconds) and INSTR SWITCH (inside normal FOV).

When conducting a light test, during which the system will be inhibited, both bulbs of each caution light should be carefully checked. The caution lights are keyed to prevent them from being replaced incorrectly, but may be interchanged with others of the same caption.

> Red lights - Warning - indicate a critical condition and require immediate action.
> Amber lights - Caution - require timely corrective action.
> Blue lights - Advisory - e.g. valve positions and unless bright blue, i.e. a valve/switch disagreement, do not require crew action.
> Green lights - Satisfactory - indicate a satisfactory or on condition.

Aural Warning Indication

The flight compartment aural warning system supplies audio signals to alert the flight crew of incorrect airplane system conditions. It includes the fire bell, take-off configuration warning, cabin altitude, landing gear configuration warning, mach/airspeed overspeed, stall warning, GPWS and TCAS. External aural warnings are: The fire bell in the wheel well and the ground call horn in the nose wheel-well for an E & E bay overheat or IRS's on DC. Only certain warnings can be silenced whilst the condition exists.

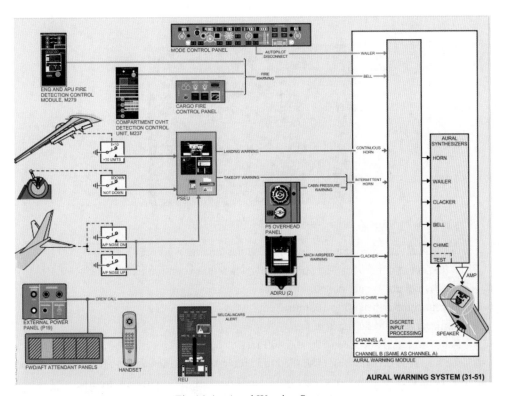

Fig.14-4 Aural Warning System

The fire bell sound is generated when a fire is detected in:
> Either engine.
> APU.
> Either forward or aft cargo compartment.
> The wheel well.

The aural warning module gives the sound of a continuous horn for a landing warning. The fire bell sound cycles three-fourths of a second on, followed by ten second off for a fire warning. The fire bell can be cancelled either by pressing master warning light or by arming the fire extinguishing system related to the fire source. For wheel-well fires during flight, the warning will be cancelled if the landing gear is down and locked. At takeoff, the fire bell is inhibited at nose wheel liftoff for a period of 20 seconds or to a latitude of 400 feet.

Fig.14-5　TCAS Display Integrated on ND

Fig.14-6　Stall Warning Panel

There are four sets of conditions which cause the landing warning.

In the first set of conditions, the horn sounds when these conditions are true:

➤ Gear is not down and locked.

➤ Flap position is from 0 to 10 units.

➤ Thrust levers are set for landing.

➤ Radio altitude is between 200 and 800 feet.

For the first set of conditions, push the horn cutout switch near the flap lever to stop the horn.

In the second set of conditions, the horn sounds when these conditions are true:

➤ Gear is not down and locked.

➤ Flap position is from 0 to 10 units.

➤ Thrust levers are set for landing.

➤ Radio altitude is less than 200 feet.

For the second set of conditions, the pilot cannot stop the horn.

In the third set of conditions, the horn sounds when these conditions are true:

➤ Gear is not down and locked.

➤ Flap position is from 15 to 25 units.

➤ Thrust levers are set for landing.

For this set of conditions, the pilot cannot stop the horn.

In the fourth set of conditions, the horn sounds when the gear is not down and locked and the flap position is more than 25 units. The pilot cannot stop the horn.

When the airplane is in the fourth set of conditions, the system inhibits the landing warning horn during a go-around. The system inhibits the warning for 12 seconds after the pilot puts the gear lever in the up position.

The aural warning module gives the sound of a fire bell for a fire warning.

These are the conditions which cause the fire warning:

➢ Engine fire.

➢ Cargo smoke.

➢ Wheel well fire.

➢ APU fire.

The fire bell stops when you push the bell cutout switch or the fire warning light.

The aural warning module gives the sound of a wailer for an autopilot disconnect warning. Do one of these things to stop the wailer:

➢ Push the autopilot disconnect switch.

➢ Push the autopilot P/RST switch.

➢ Engage the autopilot again.

The aural warning module gives the sound of a clacker for an airplane overspeed warning.

The aural warning module gives the sound of an intermittent horn for a cabin pressure warning.

Push the altitude horn cutout switch on the P5 overhead panel to stop the intermittent horn.

The aural warning unit gives the sound of a single high chime for a flight crew call. When the flight attendant calls the flight crew, there is a single high chime.

When the ground station uses SELCAL to call the flight crew, there is a single high/low chime.

The aural test checks the aural warning module for correct operation. You hear the intermittent horn and the clacker sounds for a good test. A test switch is on the top of the aural warning module. The test switch is spring loaded so that it returns to center when it is released. Use a screwdriver to turn the switch. Turn the switch clockwise to test channel A. Turn the switch counterclockwise to test channel B. Turn the switch to hear the intermittent horn. Release the switch to hear the clacker. The clacker stops at the end of the test.

Abbreviations & Acronyms

1. International Council of the Aeronautical Sciences(ICAS) 国际航空科学委员会
2. Configuration(CONFIG) 配置
3. Non-Normal Checklist (NNC) 非正常检查单
4. Ground Proximity Warning System (GPWS) 近地警告系统
5. Press to Reset(P/RST) 按压复位键
6. Selective Calling System(SELCAL) 选择性呼叫系统
7. Inertial Reference System(IRS) 惯性基准系统
8. Flight Management Computer(FMC) 飞行管理计算机
9. Reporting (RPTG) 报告
10. Transponder (XPDR) 应答机
11. Traffic Advisory /Resolution Advisory (TA/RA) 交通咨询
12. Traffic Collision Avoidance System (TCAS) 空中防撞系统
13. Air Traffic Control (ATC) 空中交通管制

14. Heater (HTR)   加热器

Words & Expressions

1. abnormal [æb'nɔːml] *adj.*   不正常的
2. configuration [kən,fɪɡə'reɪʃn] *n.*   配置
3. sensor ['sensə(r)] *n.*   传感器
4. avionic [,eɪvɪ'ɒnɪk] *adj.*   航空电子学的
5. dedicated ['dedɪkeɪtɪd] *adj.*   专用的
6. deck [dek] *n.*   舱面
7. proximity [prɒk'sɪməti] *n.*   接近
8. dual ['djuːəl] *adj.*   双重的
9. lamp [læmp] *n.*   灯

8 单词朗读

10. airliner ['eəlaɪnə(r)] *n.*   大型客机
11. annunciator [ə'nʌnsieɪtə(r)] *n.*   信号器
12. orientation [,ɔːriən'teɪʃn] *n.*   方向
13. dispatch [dɪ'spætʃ] *n./v.*   派遣
14. verify ['verɪfaɪ] *v.*   验证
15. deviation [,diːvi'eɪʃn] *n.*   偏差

9 术语专讲：复飞
（微课）

16. equivalent [ɪ'kwɪvələnt] *adj.*   相等的
17. checklist ['tʃeklɪst] *n.*   清单
18. cancel ['kænsl] *n./v.*   取消
19. encounter [ɪn'kaʊntə(r)] *v.*   遇到
20. drill [drɪl] *n.*   训练
21. inhibit [ɪn'hɪbɪt] *v.*   抑制；阻止
22. bulb [bʌlb] *n.*   电灯泡
23. interchange[ɪntə'tʃeɪndʒ] *v.*   交换
24. caption ['kæpʃn] *n.*   说明文字
25. advisory [əd'vaɪzəri] *n.*   公告
26. disagreement [,dɪsə'ɡriːmənt] *n.*   不一致
27. mach [mɑːk] *n.*   马赫
28. stall [stɔːl] *n.*   失速
29. horn [hɔːn] *n.*   喇叭
30. whilst [waɪlst] *conj.*   同时
31. cycle ['saɪkl] *v.*   循环
32. liftoff ['lɪft,ɒf] *n.*   （火箭等）发射，起飞
33. latitude ['lætɪtjuːd] *n.*   纬度
34. lever ['liːvə(r)] *n.*   控制杆
35. go-around   复飞
36. wailer ['weɪlər] *n.*   尖叫声，呼啸声
37. engage [ɪn'ɡeɪdʒ] *v.*   啮合
38. clacker [krækə] *n.*   咔嗒声
39. intermittent [,ɪntə'mɪtənt] *adj.*   间歇的，断断续续的
40. chime [tʃaɪm] *n.*   谐音
41. release [rɪ'liːs] *v.*   松开
42. screwdriver ['skruːdraɪvə(r)] *n.*   螺丝刀

43. clockwise ['klɒkwaɪz] *adj.*   顺时针方向的
44. counterclockwise[ˌkaʊntə'klɒkwaɪz] *adj.*   逆时针方向的

Exercises

10 答案

2.1 Choose the best answer from A, B, C, D options.

1. chime
   A. 烟囱   B. 化学   C. 谐音   D. 主席
2. configuration
   A. 形成   B. 配置   C. 信心   D. 轮廓
3. go-around
   A. 闲逛   B. 围绕   C. 环状   D. 复飞
4. intermittent
   A. 间歇的   B. 中断   C. 连续的   D. 间歇性
5. liftoff
   A. 下降   B. 举起   C. 脱落   D. 起飞
6. 自动驾驶
   A. autopilot   B. auto throttle   C. autobrake   D. automation
7. 放行
   A. airworthiness B. dispatch   C. deviation   D. configuration
8. 逆时针方向的
   A. clockwise   B. edelweiss   C. counterclockwise D. crosswise
9. 无线电高度
   A. radio altitude B. radio latitude C. audio altitude D. audio latitude
10. 近地警告系统
    A. Selective Calling System   B. Inertial Reference System
    C. Traffic Collision Avoidance System D. Ground Proximity Warning System

2.2 Find the best option to paraphrase the sentence.

1. The warning system provides the flight crew visual and aural indication of abnormal airplane systems conditions.
   A. The warning system supplies the flight crew with visual and aural indication of abnormal airplane systems conditions.
   B. The warning system offers the flight attendants to visual and aural indication of abnormal airplane systems conditions.
   C. The navigation system provides the flight crew visual and aural indication of normal airplane systems conditions.
   D. The lighting system provides the flight crew for visual and aural indication of abnormal airplane systems conditions.

2. The master caution system was developed for the 737 to ease pilot workload.
   A. The master caution system was developed for the 737 to erase pilot workload.
   B. The master caution system was developing for the 737 to make pilot workload heavier.
   C. The master caution system was developed for the 737 to reduce pilot workload.
   D. The monitor caution system was developed for the 737 to ease pilot workload.

3. On the ground, the master caution system will also tell you if the condition is dispatchable.
   A. On the ground, the master caution system will also tell you if the condition is dispensable.

209

B. On the ground, the master caution system will also tell you whether the condition is dispatchable.

C. On the ground, the master caution system will also tell you if the condition is airworthy.

D. On the ground, the navigation system will also tell you unless the condition is dispatchable.

4. When the airplane is in the fourth set of conditions, the system inhibits the landing warning horn during a go-around.

A. When the airplane is in the fourth set of conditions, the system protects the landing warning horn during a go-around.

B. When the airplane is in the fourth set of conditions, the system triggers the landing warning horn during a go-around.

C. When the airplane is in the fourth set of conditions, the system stops the landing warning chime during a go-around.

D. When the airplane is in the fourth set of conditions, the system restrains the landing warning horn during a go-around.

5. The aural test checks the aural warning module for correct operation.

A. The aural test checks the aural warning model for incorrect operation.

B. The aural test checks the aural warning module if it operates correctly.

C. The visual test checks the aural warning module for correct operation.

D. The aural test checks the visual warning module for correct operation.

2.3 Change the following components specified in lower case and translate them into Chinese.

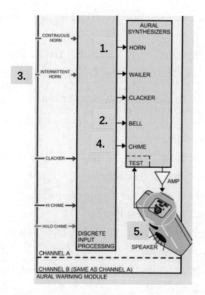

1.	
2.	
3.	
4.	
5.	

Section 3 Aviation Translation: translate the following sentences in Chinese

1. WARNING: OBEY THE INSTRUCTIONS IN THE PROCEDURE TO CLOSE THE THRUST REVERSERS. IF YOU DO NOT OBEY THE INSTRUCTIONS, INJURIES TO PERSONS AND DAMAGES TO EQUIPMENT CAN OCCUR.

11 一名航修 "老兵" 的依依深情（文本）

2. WARNING: MAKE SURE THERE ARE CAPS OR SHUNT PLUGS INSTALLED ON THE SQUIBS WHEN YOU TOUCH OR MOVE THE SQUIBS. THE SQUIB IS AN EXPLOSIVE DEVICE AND CAN CAUSE INJURIES IF IT IS ACCIDENTALLY FIRED.

Section 4 Aviation Writing

Liu Shuai is an aircraft mechanic. He is inspecting the plane and finds some faults.

Please help him write down fault descriptions. Some related words, phrases & terms and key sentences are offered as follows.

Key words, phrases & terms:

1. 为了排故   for troubleshooting
2. 液压过热警告电门   hydraulic overheat warning switch
3. 警告灯   warning light
4. 着陆警告喇叭   landing warning horn
5. 滑油滤旁通警告电门   oil filter bypass warning switch
6. 火警灯   fire warning light
7. 假警报   false alarm

Key sentences:

1. 关车时左发滑油压力警告系统瞬时闪亮（红色）。

   The oil pressure warning system of the L/H engine flashed(red light) when shutting down the engines.

2. A/T 琥珀色警告信号牌灯闪亮。

   A/T amber warning annunciator light flashed.

3. 右 PACK 灯再现时亮，警告信号抑制后熄灭。

   Right PACK light comes on during recalling and goes off after warning signal inhibited.

4. 过站检查再现故障时左组件 PACK 灯点亮，按压主警告可以熄灭。

   The L/H PACK light comes on when recalling the fault in TR, but it could extinguish by pressing master caution light.

Task: Liu Shuai has three writing tasks. The first one is that the oil pressure warning system of the right engine flashed when the engines shut down, the second is that left PACK light shows during recalling and goes off after warning signal inhibited, and the last one is that left PACK light shows when recalling the fault in transformer rectifier, but it could extinguish by pressing master caution light. Please finish the writing tasks for him.

Warning Case

Air Collision-Uberingen Disaster

12 警示案例——
乌柏林根空难

It's about Russia Bashklian Airline 2937 and DHL Express 611. Two planes collided in midair in the Swiss city of Uberlingen on July 1st, 2002. A total of 71 people, including the crew of both sides, were killed.

Warning Tips: The two flights are flying at the same altitude of about 11,000 meters in conflicting level. If the pilots of the two flights follow their traffic alert and TCAS, the disaster can be avoided.

Appendix I   Frequently-used Tools in Chinese & English

Chinese	English	Picture
扳手 / 套筒	Wrenches/sockets	
梅花扳手	Box-end wrenches	
组合扳手	Combination wrenches	
开口扳手	Open end wrenches	
棘轮扳手	Ratcheting box-end wrenches	
梅花套筒	Twelve point socket	
转接头	Adaptor	
万向转接头	Universal joint	
套筒接杆	Socket extension	
快速手柄	Speed handle	

Chinese	English	Picture
万向套筒	Universal socket	
活动扳手	Adjustable wrenches	
六角套筒	Six point socket	
力矩扳手	Torque wrenches	
螺丝刀	**Screwdriver**	
快速螺丝刀	Fast screwdriver	
十字头螺丝刀	Phillips screwdrivers	
棘轮螺丝刀	Ratcheting screwdriver	
磁性螺丝刀	**Magnetic screwdriver**	
一字头螺丝刀	Flat tip screwdriver	
偏置螺丝刀	Offset screwdriver	
摇把螺丝刀	Speeder screwdriver	

续表

Chinese	English	Picture
仪表螺丝刀	Electronic miniature screwdrivers	
钳	Pliers	
卡环钳	Convertible snap ring pliers	
尖嘴钳	Needle nose pliers	
大力钳	Locking pliers	
保险丝钳	Safety wire twisters	
鹰嘴钳	Adjustable joint pliers	
插头钳	Connector plugs	
管钳	Pipe wrenches	

Chinese	English	Picture
剥线钳	Wire strippers	
压线钳	Wire crimpers	
剪线钳	Diagonal cutters	
鱼口钳	Combination slip-joint pliers	
电气工具	Electronic tools	
手电筒	Flashing light	
对讲机	Interphone	
电烙铁	Electronic iron	
耳机	Earphone	
防静电手腕	Static wrist	

续表

Chinese	English	Picture
电枪	Electronic runner	
电钻	Electronic drills	
指挥棒	Flashing emergency light	
测量工具	Measuring tools	
钢板尺	Steel ruler	
螺旋卡尺	Outside micrometer	
组合尺	Combination set	
塞尺	Feeler (thickness) gauges	
划针	Scriber	

飞机维修专业英语——飞机系统（第二版）

Chinese	English	Picture
卡规	Caliper	
中心规	Center head	
游标卡尺	Dial caliper	
万用表	Multimeter	
杂项	Diversiform	
放大镜	Magnifier	
划刀	Utility knife	
刮刀	Scraper knife	
剪刀	Snip (scissor)	

续表

Chinese	English	Picture
锯	Saw	
凿子	Chisels	
錾子	Punch	
三脚架	Tripod (spider)	
铜榔头	Copper hammer	
胶榔头	Rubber hammer	
锉	File	
镊子	Nipper (pliers)	
铆钉枪	Rivet gun	

Chinese	English	Picture
注油枪	Grease gun	
工具箱	Tools kit	
撬棒	Tommy bar	
工作梯	Work ladder	
千斤顶	Jack	
胶带	Adhesive tape	
焊锡	Soldering tin	

续表

Chinese	English	Picture
牵引杆	Tow bar	
发光背心	Flashing vest	
护目镜	Goggles	
安全带	Life belt(safety belt)	
气枪	Pneumatic runner	
保险丝	Lockwire	
接触器	Contactor	
继电器	Relay	

Appendix II   Organizations & Related Terms in Aircraft M & O

Maintenance & Engineering Div.	机务工程部
*** Aircraft Maintenance & Overhaul Base	*** 维修基地
*** Aircraft Maintenance Co.	*** 飞机维修有限公司
Line Maintenance Dept.	航线部
Production Plan Sect.	生产室
Technical Support Sect.	技术支援室
General Affairs Sect.	综合办公室
Quality Control Sect.	质控室
Transit Shop	过站车间
Overhaul Dept.	大修部
Quality Control Sect.	质控科
Inspection Sect.	检验科
Production & Planning Sect.	生产计划科
System Shop	系统车间
Engine Shop	发动机车间
Cabin Refurbishment Shop	客舱整新车间
Structure & Machining Shop	结构机加车间
Component Repair Dept.	附件修理部
General Affairs Sect.	综合业务科
Production Sect.	生产科
Quality Control Sect.	质控科
Joint Venture Marketing	合资市场科
JV Planning Sect.	合资企划室
Electro-Mechanical Shop	机电车间
Avionic Shop	电子车间
Landing Gear Shop	起落架车间
Production Support Dept.	生产支援部
Material Management Dept.	航材管理部
Technical Management Div.	技术管理处
Technical Support	技术支援
Engineering Management	工程管理
Service Bulletin(SB)	服务通告
Airworthiness Directive(AD)	适航指令
CAD	中国适航指令

FAD	美国适航指令
EASA AD	欧洲适航指令
Production & Marketing Management Div.	生产经营处
Contract Management	合同管理
Production Planning	生产计划
Maintenance Task Operation Program(MTOP)	维修任务操作方案
Maintenance Control Center(MCC)	维修控制中心
Quality Management Div.	质量管理处
Airworthiness Management	适航管理
Flight Safety	飞行安全
Financial Div.	财务处
Human Resource Div.	人力资源处
General Office	办公室

Appendix III   Examples of AMM-PP

Lesson 1 Air Conditioning and Pressurization System

Section 5 Aviation Application: Read the following job cards with what you have learned.

AIR CONDITIONING - GENERAL - MAINTENANCE PRACTICES

1. General

A. This procedure has these tasks:

(1) Supply Conditioned Air to the Airplane

(2) Remove Conditioned Air from the Airplane

(3) Supply Conditioned Air with a Cooling Pack

(4) Remove Conditioned Air Supplied by a Cooling Pack

(5) Supply Conditioned Air with a Ground Air Source

(6) Remove Conditioned Air Supplied by a Ground Air Source

13 PP 译文   14 中国民航局《航空维修技术英语》（微课）

B. It is recommended that the conditioned air to cool the airplane on the ground be supplied from a ground air source, when practical, as an alternative to operating the cooling pack (SL 737-21-053).

TASK 21-00-00-800-801

2. Supply Conditioned Air to the Airplane (Figure 201)

A. References.

Reference	Title
SL 737-21-053	Service Letter

B. General.

SUBTASK 21-00-00-860-111

It is recommended that the conditioned air to cool the airplane on the ground be supplied from a ground air source, when practical, as an alternative to operating the cooling pack (SL737-21-053). To supply conditioned air, do this task: Supply Conditioned Air with a Ground Air Source, TASK 21-00-00-800-805.

C. Procedure.

SUBTASK 21-00-00-860-001

Do one of these tasks to supply conditioned air to the airplane:

a. Do this task: Supply Conditioned Air with a Cooling Pack, TASK 21-00-00-800-803.

b. Do this task: Supply Conditioned Air with a Ground Air Source, TASK 21-00-00-800-805.

CABIN TEMPERATURE CONTROLLER (CTC) - REMOVAL/INSTALLATION

1. General

This procedure has these tasks:

(1) A Removal of the Cabin Temperature Controller (CTC)

(2) An Installation of the Cabin Temperature Controller (CTC).

2. Cabin Temperature Controller (CTC) Removal

A. References.

Reference	Title
20-10-07-400-802	Printed Circuit Card Installation (P/B 201)

B. Location Zones.

Zone   Area

118   Electrical and Electronics Compartment - Right

C. Access Panels.

Number   Name/Location

117A   Electronic Equipment Access Door

D. Procedure.

(1) Open these circuit breakers and install safety tags: F/O Electrical System Panel, P6-4.

Row Col Number Name

A 2 C00268 AIR CONDITIONING TEMP CONTROL AUTO L

B 2 C00258 AIR CONDITIONING TEMP CONTROL AUTO R

(2) Open this access panel:

Number   Name/Location

117A   Electronic Equipment Access Door

(3) Remove the cabin temperature controller [1]. To remove the cabin temperature controller [1], do this task: Printed Circuit Card Installation, TASK 20-10-07-400-802.

3. Cabin Temperature Controller (CTC) Installation

A. References.

Reference   Title

20-10-07-000-801   E/E Box Removal (P/B 201)

24-22-00-860-811   Supply Electrical Power (P/B 201)

B. Expendables/Parts.

AMM Item Description AIPC Reference AIPC Effectivity

Controller 21-61-11-01-005 GUN 101-131

C. Location Zones.

Zone   Area

118   Electrical and Electronics Compartment - Right

D. Access Panels.

Number   Name/Location

117A   Electronic Equipment Access Door

E. Cabin Temperature Controller (CTC) Installation.

(1) Install the cabin temperature controller [1]. To install the cabin temperature controller[1], do this task: E/E Box Removal, TASK 20-10-07-000-801.

(2) Remove the safety tags and close these circuit breakers: F/O Electrical System Panel, P6-4.

Row Col Number Name.

A 2 C00268 AIR CONDITIONING TEMP CONTROL AUTO L

B 2 C00258 AIR CONDITIONING TEMP CONTROL AUTO R

F. Cabin Temperature Controller (CTC) Installation Test.

(1) Do this task: Supply Electrical Power, TASK 24-22-00-860-811.

(2) Do the test instructions that are on the front of the controller.

G. Put the Airplane Back to Its Usual Condition.

Lesson 2 Flight Control System

Section 5 Aviation Application: Read the following job cards with what you have learned.

AILERON AND AILERON TRIM CONTROL SYSTEM - MAINTENANCE PRACTICES

1. General

This procedure has these tasks:

(1) A Removal Pressure from the Aileron Hydraulic Systems A and B.

(2) Put the Aileron Hydraulic Systems A and B Back to the Condition Before Pressure Removal.

13 PP 译文   14 怎样获取 FAA 飞机维修执照（微课）

TASK 27-11-00-860-801

2. Remove Pressure from the Aileron Hydraulic Systems A and B(Figure 201)

A. General.

This procedure removes pressure from the aileron hydraulic systems A and B before you do maintenance on the aileron control system.

B. Tools/Equipment.

NOTE: When more than one tool part number is listed under the same "Reference" number, the tools shown are alternates to each other within the same airplane series. Tool part numbers that are replaced or non-procurable are preceded by "Opt:", which stands for optional.

Reference Description

SPL-1675 Assembly - Lock, Flight Controls

(Part #: F80049-12. Supplier: 81205. A/P Effectivity: 737-100, -200, -200C, -300, -400, -500, -600, -700, -700C, -700QC, -800, -900, -BBJ)

C. Location Zones.

Zone   Area

133   Main Landing Gear Wheel Well, Body Station 663.75 to Body Station 727.00 - Left

134   Main Landing Gear Wheel Well, Body Station 663.75 to Body Station 727.00 - Right

211   Flight Compartment - Left

212   Flight Compartment - Right

D. Remove Pressure from the Aileron Hydraulic Systems A and B.

WARNING: YOU MUST REMOVE PRESSURE FROM THE AILERON HYDRAULIC SYSTEMS A AND B BEFORE YOU DO MAINTENANCE ON THE AILERON CONTROL SYSTEM. THIS IS TO PREVENT INJURY TO PERSONS, AND DAMAGE TO THE EQUIPMENT BECAUSE OF ACCIDENTAL OPERATION OF THE AILERONS.

SUBTASK 27-11-00-860-001

(1) Remove pressure from the aileron hydraulic systems A and B:

a. Set the FLT CONTROL A and B switches on the forward overhead panel, P5, to the OFF position.

b. Open these circuit breakers and install safety tags:F/O Electrical System Panel, P6-2.

Row Col Number Name

B 13 C00363 FLIGHT CONTROL SHUTOFF VALVES FLT CONT

C 13 C01074 FLIGHT CONTROL SHUTOFF VALVES FLT CONT

c. Set the ELEC 2, HYD PUMPS A or ELEC 1, HYD PUMPS B switches on the forward overhead panel, P5, to the OFF position.

d. Operate the rudder pedals to decrease the remaining hydraulic system pressure.

SUBTASK 27-11-00-480-001

(2) Install the lock, SPL-1675 on the flight controls shutoff valves for systems A and B:

a. Disconnect the electrical connectors from the shutoff valves.

NOTE: The flight controls shutoff valve is part of the flight controls hydraulic module. The hydraulic module for hydraulic system A is found in the left main wheel well. The hydraulic module for hydraulic system B is found in the right main wheel well.

b. Make sure the manual override levers on the shutoff valves moved to position 2.

c. Install the lock, SPL-1675 on each shutoff valve.

Make sure to install the lockpins.

AUTOSLAT CONTROL VALVE - REMOVAL

1. General

A. This procedure has these tasks:

(1) Removal of the Autoslat Control Valve.

(2) Installation of the Autoslat Control Valve.

B. There are solenoids installed on the autoslat control valve. It is not necessary to remove the solenoids to remove the control valve.

TASK 27-83-21-000-801

2. Autoslat Control Valve Removal(Figure 401)

A. References.

Reference	Title
29-11-00-860-805	Hydraulic System A or B Power Removal (P/B 201)
32-00-01-480-801	Landing Gear Downlock Pins Installation (P/B 201)

B. Location Zones.

Zone	Area
192	Lower Wing-To-Body Fairing-Under Wing Box

C. Access Panels.

Number	Name/Location
192BR	ECS Ram Air Inlet Mixing Duct Panel - Forward

D. Prepare for the Removal.

SUBTASK 27-83-21-480-001

WARNING: MAKE SURE THE DOWNLOCK PINS ARE INSTALLED ON ALL THE LANDING GEAR. WITHOUT THE DOWNLOCK PINS, THE LANDING GEAR COULD RETRACT AND CAUSE INJURIES TO PERSONS AND DAMAGE TO EQUIPMENT.

(1) If the downlock pins are not installed on the main and nose landing gear, do this task: Landing Gear Downlock Pins Installation, TASK 32-00-01-480-801.

SUBTASK 27-83-21-860-001

(2) Do this task: Hydraulic System A or B Power Removal, TASK 29-11-00-860-805.

SUBTASK 27-83-21-010-001

(3) To gain access to the autoslat control valve, Remove this access panel:

Number	Name/Location
192BR	ECS Ram Air Inlet Mixing Duct Panel - Forward

E. Autoslat Control Valve Removal.

SUBTASK 27-83-21-860-002

(1) Open these circuit breakers and install safety tags: F/O Electrical System Panel, P6-2.

Row Col Number Name

B 14 C01070 FLIGHT CONTROL AUTOSLAT DC2

C 14 C01068 FLIGHT CONTROL AUTOSLAT DC1

SUBTASK 27-83-21-020-001

(2) Disconnect the electrical connectors [6] and [7] from the autoslat control valve [1].

SUBTASK 27-83-21-480-002

(3) Put a bucket or other suitable container to collect hydraulic fluid under the autoslat control valve [1].

SUBTASK 27-83-21-020-002

(4) Disconnect the hydraulic lines [4], [5], [8] and [9] from the autoslat control valve [1].

SUBTASK 27-83-21-020-003

(5) Remove the bolts [2] and washers [3] and remove the autoslat control valve [1].

SUBTASK 27-83-21-480-003

(6) Install caps on the hydraulic lines [4], [5], [8] and [9].

SUBTASK 27-83-21-480-004

(7) Install plugs in the hydraulic ports on the autoslat control valve [1].

Lesson 3 Fuel System

Section 5 Aviation Application: Read the following job cards with what you have learned.

13 PP 译文

14 机务常用英文技术资料（微课）

FUEL NOZZLE - INSPECTION/CHECK

1. General

This procedure has the task to inspect the fuel nozzle. Ten fuel nozzles are installed on the combustor housing.

TASK 49-31-14-200-801

2. Fuel Nozzle Inspection

A. References.

Reference   Title

49-31-14-000-801   Fuel Nozzle Removal (P/B 401)

49-31-14-400-801   Fuel Nozzle Installation (P/B 401)

B. Location Zones.

Zone   Area

211   Flight Compartment - Left

315   APU Compartment - Left

316   APU Compartment - Right

C. Procedure.

SUBTASK 49-31-14-020-001

(1) Do this task: Fuel Nozzle Removal, TASK 49-31-14-000-801.

SUBTASK 49-31-14-210-001

(2) Do these steps to inspect the fuel nozzle:

a. Examine the two inlet fittings for the fuel nozzle for crossed, stripped and peened threads.

b. Examine the mounting flange for burrs and nicks.

c. Examine the fuel nozzle for burrs, nicks, scratches and carbon buildup.

d. Examine the fuel nozzle for blockage.

e. Examine the locating pin for damage.

f. If you find any of the above damage, replace the fuel nozzle.

SUBTASK 49-31-14-420-001

(3) Do this task: Fuel Nozzle Installation, TASK 49-31-14-400-801.

SYSTEM TEST - FUEL QUANTITY INDICATING SYSTEM

TASK 28-41-00-730-801

A. General.

The fuel quantity processor unit, M121, will do a capacitance test of the tank units and compensators.

NOTE: This test is a dry capacitance test (no fuel in the tank).

B. References.

Reference	Title
12-11-00-650-802	Pressure Refuel Procedure (P/B 301)
12-11-00-650-804	Drain the Fuel from the Sumps after Defueling (P/B 301)
28-26-00-650-801	Fuel Tank Defueling (P/B 201)
28-26-00-650-802	Tank to Tank Fuel Transfer (P/B 201)
28-41-21-710-801	Tank and Compensator Units Test (P/B 501)
FIM 28-41	TASK 801 FQIS BITE Procedure

C. Dry Capacitance Test in the Main Equipment Center (No Fuel in the Tank).

(1) Do this task: Operational Test - Fuel Quantity Indicating System, TASK 28-41-00-710-801.

a. If the BITE test shows faults, then do the corrective action for the fault shown.

b. If the BITE test does not show faults, then continue.

(2) Defuel the applicable fuel tank (TASK 28-26-00-650-801) or transfer fuel out of the applicable tank (TASK 28-26-00-650-802).

(3) Drain the fuel from the applicable fuel tank (TASK 12-11-00-650-804).

(4) On the Control Display Unit (CDU),push the line select key next to the MAINT prompt on the CDU Initial Reference Index.

NOTE: This brings you to the MAINT BITE INDEX.

a. Push the line select key next to the FQIS prompt.

b. Push the line select key next to the INPUT MONITORING prompt.

c. Push the NEXT key to get to the second INPUT MONITORING page.

d. Make sure the quantities shown under TANK UNIT CAP are in this range for the applicable tank:

➢ No. 1 Tank and No. 2 Tank: 399.40 pf maximum/390.30 pf minimum.

➢ Center Tank: 574.50 pf maximum/563.00 pf minimum.

e. Make sure the quantities shown under COMPENSATOR CAP are in this range for the applicable tank:

59.2 pf maximum/57.6 pf minimum.

NOTE: These values apply to the No. 1, No. 2, and the center tank.

(5) If the capacitances are not in the correct range for empty tank capacitances, do this task: FIM 28-41 TASK 801.

a. If the BITE test shows faults, then do the corrective action for the fault shown.

b. If the BITE test does not show faults, then do the tank unit and compensator test from the wing spar (TASK 28-41-21-710-801) to isolate the problem.

6. If no more tests are necessary, refuel the applicable tank.

Lesson 4 Hydraulic System

Section 5 Aviation Application: Read the following job cards with what you have learned.

13 PP 译文   14 ATA-100 介绍
（微课）

HYDRAULIC POWER - CLEANING/PAINTING

1. General

A. This procedure has a task to apply corrosion protection compounds to components of the hydraulic system.

B. The high pressure hydraulic lines are unpainted corrosion resistant steel (CRES). The low pressure hydraulic lines are 5000 or 6000 series aluminum alloys. Valves and fittings are either andozied aluminum or CRES. Service experience has shown that these components are relatively corrosion free unless they are exposed to severe operating environments.

C. Clamps are usually manufactured from solid nylon or silicon rubber cushioned steel.

TASK 29-00-00-916-801

2. Hydraulic Components - Corrosion Protection

A. References.

Reference	Title
51-21-91-620-802	Application of Corrosion Inhibiting Compound (P/B 701)
SL SL-29-37	Service Letter
SRM 51-10-02	Structural Repair Manual

B. Consumable Materials.

Reference Description Specification

A00436 Sealant - Fuel Tank BMS5-45

B00148 Solvent - Methyl Ethyl Ketone (MEK) ASTM D740

G00009 Compound - Organic Corrosion Inhibiting BMS3-23

G00034 Cotton Wiper - Process Cleaning Absorbent Wiper (Cheesecloth, Gauze)BMS15-5

C. Hydraulic Lines and Fittings - Corrosion Protection.

SUBTASK 29-00-00-916-001

(1) For periodic inspections of the hydraulic lines, do the steps that follow:

a. Examine the hydraulic lines for signs of corrosion.

NOTE: Corrosion on aluminum components is white in color. Corrosion on CRES components is black in color.

b. If clamps are removed, do an inspection of the tubing for signs of corrosion on the surface where the clamp touches the tubing.

c. If corrosion is found, remove the corrosion: SRM 51-10-02.

CAUTION: DO NOT APPLY CORROSION INHIBITING COMPOUND, G00009 TO SILICONE RUBBER, OR RUBBER CLAMP CUSHIONS. IT CAN CAUSE DAMAGE TO SILICONE RUBBER.

d. Use a cotton wiper, G00034 moist with corrosion inhibiting compound, G00009 to apply a thin layer of the compound to the hydraulic line. Application of Corrosion Inhibiting Compound, TASK 51-21-91-620-802.

SUBTASK 29-00-00-916-002

(2) If the hydraulic lines have been cleaned with steam or high pressure water and detergent, do the steps that follow:

a. Remove unwanted moisture with a cotton wiper, G0003.

b. Use a cotton wiper, G00034 moist with corrosion inhibiting compound, G00009 to apply a thin layer of the compound to the hydraulic line. Application of Corrosion Inhibiting Compound, TASK 51-21-91-620-802.

SUBTASK 29-00-00-916-003

(3) If the hydraulic lines get a scratch or gouge during maintenance or repair work, do the step that follows:

Use a cotton wiper, G00034 moist with corrosion inhibiting compound, G00009 to apply a thin layer of the compound to the hydraulic line. Application of Corrosion Inhibiting Compound, TASK 51-21-91-620-802.

D. Improved Corrosion Protection.

NOTE: SL SL-29-37 gives corrosion protection for external threads and cavities of hydraulic actuators and components that are open to moisture. Batco 8401 Number 1 grease plus a bead of BMS 5-45, Type II, Class B-1/2, optional Class 3-2 sealant are applied. This grease was selected because of compatibility with O-ring seal material and fair resistance to BMS 3-11 hydraulic fluid.

SUBTASK 29-00-00-916-004

For applying the corrosion protection do the steps that follow:

a. Apply a thin layer of the grease to the faying surfaces of threads or flanges.

b. Assemble the component and clean off unwanted grease.

c. Clean the areas where sealant will be applied with solvent, B00148 or equivalent.

d. Apply a bead of sealant, A00436 to the joints that were greased.

e. Allow the sealant to cure 48 h and make sure that it has adhered to the surfaces.

PRESSURE RELIEF VALVE - INSTALLATION

TASK 29-09-05-400-801

A. References.

Reference　　　　　　Title

29-09-00-860-802　　Hydraulic Reservoirs Depressurization (P/B 201)

B. Consumable Materials.

Reference Description Specification

D00054 Fluid - Hydraulic Assembly Lubricant - MCS 352B (Formerly Monsanto MCS 352B)

D00153 Fluid - Hydraulic Fluid, Fire Resistant (Interchangable And Intermixable With BMS 3-11 Type V)BMS3-11 Type IV

C. Location Zones.

Zone　Area

134　　Main Landing Gear Wheel Well, Body Station 663.75 to Body Station 727.00 - Right

D. Pressure Relief Valve Installation.

(1) Remove the plugs/caps from the tube ends and pressure relief valve ports.

(2) Lightly lubricate the new packings (o-rings) with MCS 352B fluid, D00054 or hydraulic fluid, D00153.

(3) Install the packings (o-rings) and the unions to the pressure relief valve ports, and tighten.

(4) Install the pressure relief valve and tighten the swivel nuts.

NOTE: The "IN-OUT" flow arrow on the pressure relief valve must point away from the adjacent cross fitting assembly. The inlet side of the pressure relief valve is adjacent to the cross fitting assembly. The outlet side of the pressure relief valve is adjacent to the APU fuel line shroud drain mast tubing.

(5) Install the clamp, washers, and screws to the pressure relief valve.

E. Post-Installation Leakage Check.

(1) Pressurize the pneumatic crossover manifold duct.

(2) Use a solution of soap and water to check that no air leaks at the swivel nut connection on the inlet side of the pressure relief valve.

(3) Depressurize the pneumatic cross manifold duct.

(4) If necessary, do this task: Hydraulic Reservoirs Depressurization.

Lesson 5　Pneumatic System

Section 5　Aviation Application: Read the following job cards with what you have learned.

13 PP 译文　　14 飞机维修手册
（微课）

SUPPLY PRESSURE TO THE PNEUMATIC SYSTEM WITH AN EXTERNAL GROUND AIR SOURCE

TASK 36-00-00-860-802

1. Supply Pressure to the Pneumatic System with an External Ground Air Source (Figure 201)

A. References.

Reference	Title
24-22-00-860-811	Supply Electrical Power (P/B 201)

B. Location Zones.

Zone	Area
192	Lower Wing-To-Body Fairing - Under Wing Box

C. Access Panels.

Number	Name/Location
192DR	ECS High Pressure Access Door

D.　Procedure.

SUBTASK 36-00-00-860-002

(1) Do this task: Supply Electrical Power, TASK 24-22-00-860-811.

SUBTASK 36-00-00-860-003

(2) To gain access to the ground pneumatic service connector, do this step: Open this access panel:

Number	Name/Location
192DR	ECS High Pressure Access Door

SUBTASK 36-00-00-860-004

(3) Connect the ground pneumatic service line to the ground pneumatic service connector.

SUBTASK 36-00-00-860-005

WARNING: DO NOT SUPPLY MORE THAN 60 PSI (413.7 kPa) OF PRESSURE TO THE PNEUMATIC SYSTEM. IF YOU SUPPLY TOO MUCH PRESSURE, DAMAGE TO EQUIPMENT AND INJURIES TO PERSONNEL CAN OCCUR.

(4) Start the external ground air source.

NOTE: Do not supply more than 60.0 psi (413.7 kPa) of pressure.

SUBTASK 36-00-00-860-006

(5) Put the ISOLATION VALVE switch on the P5-10, forward overhead panel to the OPEN position.

SUBTASK 36-00-00-860-007

(6) Monitor the dual duct pressure indicator on the P5-10, forward overhead panel.

Make sure that there is movement shown by both duct pressure needles and the pressures shown are normal.

NOTE: Make sure that the duct pressure needles do not show more than 60.0 psi (413.7 kPa).

GROUND PNEUMATIC CONNECTOR CHECK VALVE INSTALLATION

TASK 36-13-03-400-801 (Figure 401)

A. References.

Reference	Title
36-00-00-860-803	Supply Pressure to the Pneumatic System with the APU (P/B 201)
36-00-00-860-804	Supply Pressure to the Pneumatic System with One or Both Engines(P/B 201)

B. Location Zones.

Zone	Area
192	Lower Wing-To-Body Fairing - Under Wing Box

C. Access Panels.

Number	Name/Location
192CR	ECS Access Door
192DR	ECS High Pressure Access Door

D. Install the Check Valve.

SUBTASK 36-13-03-420-001

(1) Put the check valve [5] and lower gasket [4] in the correct position.

SUBTASK 36-13-03-420-002

(2) Install the nuts [6] and washers [7], at 12 locations.

Tighten the nuts [6] to 26-30 pound-inches (2.89-3.34 Newton-meters).

E. Check Valve Installation Test.

SUBTASK 36-13-03-790-001

Do a check for leakage of the check valve [5].

a. To use the APU to supply pressure to the pneumatic manifold, do this task: Supply Pressure to the Pneumatic System with the APU, TASK 36-00-00-860-803.

b. To use the engine to supply pressure to the pneumatic manifold, do this task: Supply Pressure to the Pneumatic System with One or Both Engines, TASK 36-00-00-860-804.

c. Examine the check valve [5] for leakage.

NOTE: Diffused leakage is permitted, concentrated leakage must be repaired.

F. Put the Airplane Back To Its Usual Condition.

SUBTASK 36-13-03-410-001

(1) Close these access panels:

Number	Name/Location
192CR	ECS Access Door
192DR	ECS High Pressure Access Door

SUBTASK 36-13-03-860-003

(2) Remove the DO-NOT-OPERATE tags from these switches on the P5-10 panel:

a. BLEED 1

b. BLEED 2

c. APU BLEED

Lesson 6 Water and Waste System

Section 5 Aviation Application: Read the following job
cards with what you have learned.

13 PP 译文   14 飞机维修手册
SDS（微课）

FORWARD LAVATORY DRAIN VALVE - MAINTENANCE PRACTICES

1.   General

A. This procedure contains scheduled maintenance task data.

B. The Forward Lavatory Drain Valve is referred to as the Drain Valve in this procedure.

C. This procedure has this task:

An operational check of the drain valve in the forward lavatory.

TASK 38-11-06-000-802-001

2. Forward Lavatory Drain Valve Operational Check (Figure 201)

A. General.

This procedure is a scheduled maintenance task.

B. References.

Reference   Title

38-42-00-800-802   Potable Water System - Pressurization (P/B 201)

C. Location Zones.

Zone   Area

200   Upper Half of Fuselage

D. Prepare for the Operational Check.

SUBTASK 38-11-06-860-025

(1) Do this task: Potable Water System - Pressurization, TASK 38-42-00-800-802.

SUBTASK 38-11-06-010-004-001

(2) Get access to the lavatory.

SUBTASK 38-11-06-010-005-001

(3) Open the cabinet below the sink to get access to the water supply shutoff valve and drain valve.

E. Drain Valve Operational Check.

SUBTASK 38-11-06-860-009-001

(1) Make sure the water supply shutoff valve is in the SUPPLY ON position.

SUBTASK 38-11-06-790-002-001

(2) Make sure the handle for the drain valve is in the closed position.

SUBTASK 38-11-06-860-010-001

(3) Open the water faucet, make sure water flows from the faucet.

NOTE: This step checks for water pressure. Water pressure is required for the drain valve
leak check.

SUBTASK 38-11-06-860-023

(4) Close the water faucet.

SUBTASK 38-11-06-790-003-001

(5) Make sure that there is no water leakage from the drain port after approximately 5 min.

F.   Put the Airplane Back in Its Usual Condition.

SUBTASK 38-11-06-010-006-001

Close the cabinet below the sink.

POTABLE WATER SYSTEM – SERVICING

1. General

This procedure has these tasks:

(1) Potable water system - drain.

(2) Potable water tank - fill.

TASK 12-14-01-600-801

2. Potable Water System - Drain(Figure 301, Figure 302)

A. Location Zones.

Zone   Area

200   Upper Half of Fuselage

B. Access Panels.

Number   Name/Location

146AR   Water Service Door

C. Drain the Potable Water System.

(1) Open this access panel:

Number   Name/Location

146AR   Water Service Door

(2) Connect a drain line to each of the drain ports.

NOTE: There are two drain port locations. The first is the forward drain port for the forward lavatory/galley. The second is the aft drain/overflow port for the water service panel. The drain ports have 1/2 - 14 ANPT threads.

(3) Make sure the shutoff valve for each lavatory and each wet galley is in the OPEN position. (Figure 302).

NOTE: The shutoff valve is found adjacent to the sink of a wet galley.

(4) Turn the handle for the water drain valve on the water service panel to open the water tank drain valve.

NOTE: This drains the potable water tank and the water system aft of the wings.

(5) Make sure the supply valve for each lavatory is in the ON position.

NOTE: The supply valve is found below the sink in the lavatory.

(6) Make sure the drain valve for forward lavatory is in the OPEN TO DRAIN position.

NOTE: The drain valve is found below the sink in the lavatory.

(7) If it is installed, open the drain valve to drain the water from each coffee maker or water boiler.

(8) Open the galley water faucet to drain the water from the galley water system.

Close the galley water faucet when the water flow stops.

(9) Make sure the potable water system is empty.

(10) If you do not fill the potable water tank immediately after you drain the system, do these steps:

Open these circuit breakers and install safety tags:

CAPT Electrical System Panel, P18-3

Row Col Number Name

F 13 C00104 LAVATORY WATER HEATER A

F 14 C01073 LAVATORY WATER HEATER D

F 15 C01096 LAVATORY WATER HEATER E GUN 001-015

F 16 C01510 LAVATORY WATER HEATER G

Power Distribution Panel Number 1, P91

Row Col Number Name   GUN ALL

A 18 C00873 POT WATER COMPRESSOR

C 9 C00138 WATER QTY IND

(11) If it is installed, move the drain valve for each coffee maker or water boiler to the CLOSED position.

(12) Move the drain valve in forward lavatory to the CLOSED position.

(13) Turn the handle for the drain valve on the water service panel to close the water tank drain valve.

(14) Disconnect the drain lines from the drain ports.

(15) Close this access panel:

Number   Name/Location

146AR   Water Service Door

Lesson 7 Engine Oil

Section 5 Aviation Application: Read the following PP material with what you have learned.

13 PP 译文　　14 飞机维修手册——PP（微课）

INTERNAL OIL SYSTEM INSPECTION

TASK 79-00-00-200-803-F00

1. Internal Oil System Inspection

A. General.

(1) This task is a borescope inspection of areas of the internal engine where oil leakage could occur. When you do the fault isolation for an oil consumption problem, an inspection for an internal oil leak could be necessary.

(2) This task includes a borescope inspection of the High Pressure Compressor (HPC).

B. References.

Reference	Title
71-00-02-000-801-F00	Power Plant Removal (P/B 401)
71-00-02-400-801-F00	Power Plant Installation (P/B 401)
71-71-00-200-801-F00	Engine Vents and Drains Inspection (P/B 601)
72-00-00-200-804-F00	Borescope Inspection of the HP Compressor Blades (P/B 601)
8-31-00-010-801-F00	Open the Thrust Reverser (Selection) (P/B 201)
78-31-00-010-804-F00	Close the Thrust Reverser (Selection) (P/B 201)

C. Location Zones.

Zone	Area
411	Engine 1 - Engine
421	Engine 2 – Engine

D. Procedure.

SUBTASK 79-00-00-010-003-F00

WARNING: DO THESE SPECIFIED TASKS IN THE CORRECT SEQUENCE

BEFORE YOU OPEN THE THRUST REVERSER: RETRACT THE LEADING EDGE, DO THE DEACTIVATION PROCEDURES FOR THE LEADING EDGE AND THE THRUST REVERSER (FOR GROUND MAINTENANCE), AND OPEN THE FAN COWL PANELS. IF YOU DO NOT OBEY THE ABOVE SEQUENCE, INJURIES TO PERSONS AND DAMAGE TO EQUIPMENT CAN OCCUR.

(1) Do this task: Open the Thrust Reverser (Selection), TASK 78-31-00-010-801-F00.

SUBTASK 79-00-00-290-001-F00

(2) To do a check of the stage 1 and 2 blades on the HPC, do this task: Borescope Inspection of the HP Compressor Blades, TASK 72-00-00-200-804-F00.

NOTE: Look for signs of oil on the blades.

If the stage 1 and 2 blades on the HPC are wet with oil and the engine oil consumption is above serviceable limits Engine Vents and Drains Inspection.

TASK 71-71-00-200-801-F00, replace the engine. These are the tasks:

➢ Power Plant Removal, TASK 71-00-02-000-801-F00

➢ Power Plant Installation, TASK 71-00-02-400-801-F00.

SUBTASK 79-00-00-410-002-F00

WARNING: OBEY THE INSTRUCTIONS IN THE PROCEDURE TO CLOSE THE THRUST REVERSERS. IF YOU DO NOT OBEY THE INSTRUCTIONS, INJURIES TO PERSONS AND DAMAGE TO EQUIPMENT CAN OCCUR.

(3) Do this task: Close the Thrust Reverser (Selection), TASK 78-31-00-010-804- F00.

OIL TANK - REPAIRS

1. General

This procedure contains one task: the replacement of the filler cap packing or the filler cap.

TASK 79-11-01-300-801-F00

2. Replacement of the Filler Cap Packing or Filler Cap (Figure 801)

A. General.

(1) Replace the filler cap packing if you find an oil leak between the oil tank and the filler cap.

(2) Replace the filler cap if you find it damaged.

(3) The oil tank is installed on the right side of the fan case.

(4) This procedure refers to the oil tank filler cap as the filler cap.

B. References.

Reference	Title
71-11-02-010-801-F00	Open the Fan Cowl Panels (P/B 201)
71-11-02-410-801-F00	Close the Fan Cowl Panels (P/B 201)

C. Consumable Materials.

Reference	Description	Specification
D00599 [CP2442]	Oil - Engine	
G00034	Cotton Wiper - Process Cleaning Absorbent Wiper (Cheesecloth, Gauze)	BMS15-5

D. Expendables/Parts.

AMM Item	Description	AIPC Reference	AIPC Effectivity
1	Packing	Not Specified	

E. Location Zones.

Zone   Area

411   Engine 1 - Engine

421   Engine 2 - Engine

F. Procedure.

SUBTASK 79-11-01-010-004-F00

(1) Do this task: Open the Fan Cowl Panels, TASK 71-11-02-010-801-F00.

SUBTASK 79-11-01-960-001-F00

(2) Do these steps to replace the packing [1] or the filler cap:

CAUTION: MAKE SURE THAT THE OIL SCUPPER IS CLEAN. IF THE OIL SCUPPER IS NOT CLEAN, CONTAMINATION OF THE OIL TANK CAN OCCUR DURING THE SAMPLING OPERATION.

a. Use a cotton wiper, G00034, to clean the oil scupper.

b. Lift the filler cap handle.

c. Turn the filler cap handle counterclockwise to open it.

d. Pull the filler cap from the gravity fill port.

e. Disconnect the from the on the filler cap

f. Remove and discard the packing [1] from the groove of the filler cap.

NOTE: If you find the filler cap is damaged, replace it.

g. Lubricate a new packing [1] with oil, D00599 [CP2442].

CAUTION: MAKE SURE YOU INSTALL THE PACKING CORRECTLY ON THE FILLER CAP. IF YOU DO NOT INSTALL THE PACKING CORRECTLY, OIL LOSS CAN OCCUR DURING ENGINE OPERATION AND CAN CAUSE DAMAGE TO THE ENGINE.

h. Install the packing [1] in the groove of the filler cap.

i. Connect the chain to the split ring on the filler cap.

j. Put the filler cap in its position in the gravity fill port.

k. Push the filler cap in and turn the handle clockwise to lock it.

l. Push the filler cap handle down to the closed position.

G. Put the Airplane Back to Its Usual Condition.

SUBTASK 79-11-01-410-005-F00

Do this task: Close the Fan Cowl Panels, TASK 71-11-02-410-801-F00.

Lesson 8 Communications System

Section 5 Aviation Application: Read the following PP material with what you have learned.

13 PP 译文   14 英文手册的警告
—告诫—注意辨析
（微课）

HF TRANSCEIVER - REMOVAL/INSTALLATION

1. General

A. This procedure has these tasks:

(1) A Removal of the HF Transceiver.

(2) An Installation of the HF Transceiver.

B. The No. 1 and No. 2 HF transceivers are located in the aft cargo compartment on the electronic equipment rack E6.

C. The HF transceiver is located in the main equipment center on the electronic equipment rack E8.

D. The No. 1 and No. 2 HF transceivers are located in the main equipment center on the electronic equipment rack E8.

TASK 23-11-21-000-801

2. HF Transceiver -Removal (Figure 401)

A. References.

Reference   Title

20-10-07-000-801   E/E Box Removal (P/B 201)

20-40-12-000-802   ESDS Handling for Metal Encased Unit Removal (P/B 201)

B. Location Zones.

Zone   Area

117   Electrical and Electronics Compartment - Left

142   Aft Cargo Compartment - Right

211   Flight Compartment - Left

212   Flight Compartment - Right

C. Access Panels.

Number   Name/Location

117A   Electronic Equipment Access Door

822   Aft Cargo Door

D. Removal Procedure.

SUBTASK 23-11-21-860-002

(1) Open these circuit breakers and install safety tags:

a. CAPT Electrical System Panel, P18-2.

Row Col Number Name

E 11 C00839 COMMUNICATIONS HF1

b. F/O Electrical System Panel, P6-1.

Row Col Number Name

D 2 C00857 COMMUNICATIONS HF 2

SUBTASK 23-11-21-010-001

(2) Open this access panel:

Number   Name/Location

117A   Electronic Equipment Access Door

SUBTASK 23-11-21-010-002

(3) Open this access panel:

Number   Name/Location

822   Aft Cargo Door

SUBTASK 23-11-21-020-001

CAUTION: DO NOT TOUCH THE CONNECTOR PINS OR OTHER CONDUCTORS ON THE HF TRANSCEIVER. IF YOU TOUCH THESE CONDUCTORS, ELECTROSTATIC DISCHARGE CAN CAUSE DAMAGE TO THE HF TRANSCEIVER.

(4) Before you touch the HF TRANSCEIVER [1], do this task: ESDS Handling for Metal Encased Unit Removal, TASK 20-40-12-000-802.

SUBTASK 23-11-21-020-002

(5) To remove the HF TRANSCEIVER [1], do this task: E/E Box Removal, TASK 20-10-07-000-801

STATIC DISCHARGER - REMOVAL

TASK 23-61-00-000-801

A. General.

(1) The static dischargers are installed on the edge of the airplane wing and tail.

(2) The discharger is held by a setscrew on its base.

(3) The base is attached to the airplane surface.

B. References.

Reference   Title

24-22-00-860-812   Remove Electrical Power (P/B 201)

27-11-00-860-801   Remove Pressure from the Aileron Hydraulic Systems A and B(P/B 201)

C. Location Zones.

Zone   Area

300   Empennage

500   Left Wing

600   Right Wing

D. Procedure.

(1) Remove pressure and power from the hydraulic systems for the applicable flight control surfaces: do this task—Remove Pressure from the Aileron Hydraulic Systems A and B, or, do this task—Remove Pressure from the Rudder Hydraulic Systems A, B, and Standby, or, do this task—Remove Pressure from the Elevator Hydraulic Systems A and B.

(2) Do this task: Remove Electrical Power.

(3) Get access to the defective static discharger.

(4) Loosen the static discharger setscrew [2].

(5) Remove the static discharger [1] from the discharger base [3].

(6) If it is necessary, put the hydraulic systems back to the usual condition: do this task—Put the Aileron Hydraulic Systems A and B Back to the Condition Before Pressure Removal, or, do this task—Put the Rudder Systems A, B, and Standby Back to the Condition Before the Pressure Removal, or, do this task—Put the Elevator Systems A and B Back to the Condition Before the Pressure Removal .

Lesson 9 Electrical Power System

Section 5 Aviation Application: Read the following PP material with what you have learned.

13 PP 译文   14 页块（微课）

EXTERNAL POWER CONTACTOR - REMOVAL/INSTALLATION

1. General

This procedure has two tasks:

(1) A Removal of the External Power Contactor.

(2) An Installation of the External Power Contactor.

TASK 24-41-12-000-801

2. External Power Contactor Removal (Figure 401)

A. General.

The External Power Contactor, C937 is located in the P92 Power Distribution Panel.

B. References.

Reference   Title

24-22-00-860-812   Remove Electrical Power (P/B 201)

C. Location Zones.

Zone   Area

117   Electrical and Electronics Compartment - Left

118   Electrical and Electronics Compartment - Right

D. Access Panels.

Number   Name/Location

114AR   External Power Receptacle Door

117A   Electronic Equipment Access Door

E. Prepare for removal.

SUBTASK 24-41-12-860-001

WARNING: REMOVE ELECTRICAL POWER BEFORE REMOVING OR INSTALLING CONTROL BREAKERS IN POWER DISTRIBUTION PANELS. HIGH VOLTAGES COULD BE FATAL.

(1) Do this task: Remove Electrical Power, TASK 24-22-00-860-812.

Make sure that all of the power warning lights on the P92 panel are off.

SUBTASK 24-41-12-930-002

(2) Open this access panel and attach a DO-NOT-OPERATE tag to External Power Receptacle:

Number   Name/Location

114AR   External Power Receptacle Door

SUBTASK 24-41-12-010-002

(3) Open this access panel to get access to the main equipment center:

Number   Name/Location

117A   Electronic Equipment Access Door

F. Procedure.

SUBTASK 24-41-12-020-001

Do these steps to remove the external power contactor [1]:

(1) Open the P92 panel for access to the contactor [1].

(2) Loosen the two screws and remove the electrical connector from the contactor [1].

(3) Remove the six bolts [2] and washers [3] that hold the contactor [1].

(4) Remove the contactor [1].

STATIC INVERTER INSTALLATION

TASK 24-34-21-400-801

A. General.

The M9 Static Inverter is located on the E2-2 equipment rack in the main equipment center.

B. References.

Reference   Title

20-10-07-400-801   E/E Box Installation (P/B 201)

20-40-12-400-802   ESDS Handling for Metal Encased Unit Installation (P/B 201)

C. Expendables/Parts.

AMM Item Description AIPC Reference AIPC Effectivity

1 Static inverter 24-34-21-01-005 GUN ALL

D. Location Zones.

Zone   Area

117   Electrical and Electronics Compartment - Left

E. Access Panels.

Number Name/Location

117A   Electronic Equipment Access Door

F. Static Inverter Installation.

WARNING: MAKE SURE THAT POWER IS STILL REMOVED FROM STATIC INVERTER WIRING BEFORE INSTALLING STATIC INVERTER. POWER PRESENT ON WIRES CAN CAUSE INJURY TO PERSONS.

CAUTION: DO NOT TOUCH THE STATIC INVERTER BEFORE YOU DO THE PROCEDURE FOR DEVICES THAT ARE SENSITIVE TO ELECTROSTATIC DISCHARGE. ELECTROSTATIC DISCHARGE CAN CAUSE DAMAGE TO THE STATIC INVERTER.

(1) Do this task: ESDS Handling for Metal Encased Unit Installation, TASK 20-40-12-400-802.

(2) Install the static inverter [1]. To install it, do this task: E/E Box Installation.

(3) Do these steps to connect the static inverter:

a. Put the wires on the terminal studs of the static inverter.

b. Install the nut [2], lockwasher [3] and washer [4] on the terminal stud.

c. Tighten the nut to 135-145 inch-pounds (15.3-16.4 Newton meters).

d. Install the nut [5], lockwasher [6] and washer [7] on the terminal stud.

e. Tighten the nut to 65-75 inch-pounds (7.3-8.5 Newton meters).

f. Do these steps to install the terminal block covers:

➢ Hold the two terminal block covers [11] in place.

➢ Install the four screws [8], lockwashers [9] and washers [10] that hold the terminal block covers.

g. Install the electrical connector on the static inverter.

Lesson 10 External Power

Section 5 Aviation Application: Read the following PP material with what you have learned.

13 PP 译文 14 计划性维护与
非计划性维护文献
（微课）

EXTERNAL POWER CONTACTOR - REMOVAL/INSTALLATION

1. General

This procedure has two tasks:

(1) A removal of the External Power Contactor.

(2) An installation of the External Power Contactor.

TASK 24-41-12-000-801

2. External Power Contactor Removal (Figure 401)

A. General.

The External Power Contactor, C937 is located in the P92 Power Distribution Panel.

B. References.

Reference   Title

24-22-00-860-812 Remove Electrical Power (P/B 201)

C. Location Zones.

Zone Area

117   Electrical and Electronics Compartment - Left

118   Electrical and Electronics Compartment - Right

D. Access Panels.

Number   Name/Location

114AR   External Power Receptacle Door

117A   Electronic Equipment Access Door

E. Prepare for removal.

SUBTASK 24-41-12-860-001

WARNING: REMOVE ELECTRICAL POWER BEFORE REMOVING OR INSTALLING CONTROL BREAKERS IN POWER DISTRIBUTION PANELS. HIGH VOLTAGES COULD BE FATAL.

(1) Do this task: Remove Electrical Power, TASK 24-22-00-860-812.

Make sure that all of the power warning lights on the P92 panel are off. SUBTASK 24-41-12-930-002

(2) Open this access panel and attach a DO-NOT-OPERATE tag to External Power Receptacle:

Number   Name/Location

114AR   External Power Receptacle Door

SUBTASK 24-41-12-010-002

(3) Open this access panel to get access to the main equipment center:

Number   Name/Location

117A   Electronic Equipment Access Door

F. Procedure

SUBTASK 24-41-12-020-001

Do these steps to remove the external power contactor [1]:

a. Open the P92 panel for access to the contactor [1].

b. Loosen the two screws and remove the electrical connector from the contactor [1].

c. Remove the six bolts [2] and washers [3] that hold the contactor [1].

d. Remove the contactor [1].

EXTERNAL POWER RECEPTACLE PIN INSPECTION

TASK 24-41-11-200-802 (Figure 601)

A. General.

(1) The External Power Receptacle is located on the lower right hand side of the airplane. It is installed forward of the nose gear wheel well.

(2) The External Power Receptacle Pin Inspection uses a GO/NO-GO gauge to make sure that the pins are not worn. If the pins are worn, the external power receptacle should be replaced.

B. References.

Reference   Title

24-22-00-860-812   Remove Electrical Power (P/B 201)

24-41-11-000-801   External Power Receptacle Removal (P/B 401)

24-41-11-400-801   External Power Receptacle Installation (P/B 401)

C. Tools/Equipment.

NOTE: When more than one tool part number is listed under the same "Reference" number, the tools shown are alternates to each other within the same airplane series. Tool part numbers that are replaced or non-procurable are preceded by "Opt:", which stands for Optional.

Reference   Description

SPL-1625   Wear Gauge Set - Ground Power Plug and Receptacle

(Part #: F70284-1, Supplier: 81205, A/P Effectivity: 737-ALL)

D. Location Zones.

Zone   Area

116   Nose Landing Gear Wheel Well - Right

E. Access Panels.

Number   Name/Location

114AR   External Power Receptacle Door

F. Procedure.

SUBTASK 24-41-11-860-007

(1) Do this task: Remove Electrical Power, TASK 24-22-00-860-812.

SUBTASK 24-41-11-010-008

(2) Open this access panel:

Number   Name/Location

114AR   External Power Receptacle Door

SUBTASK 24-41-11-010-009

(3) Remove External Power Plug from receptacle, if it is installed.

SUBTASK 24-41-11-220-001

(4) Inspect the External Power Receptacle pins for wear as follows:

SUBTASK 24-41-11-200-802

(5) Inspect the External Power Receptacle pins for wear as follows:

TASK 24-41-11-200-802

CAUTION: DO NOT USE TOO MUCH FORCE WHEN PUSHING THE WEAR GAGE ONTO THE PINS. THE WEAR GAGE IS A GO/NO GO TOOL AND SHOULD NOT FIT OVER THE PINS. THE USE OF TOO MUCH FORCE COULD CAUSE DAMAGE TO PINS.

a. Try to slide the wear gauge set, SPL-1625 over the external power receptacle pins

NOTE: The F70284-1 is a gage set. Use the -2 on the four large pins A, B, C and D. Use the-3 on the two small pins E and F.

b. Make sure the gage does not slide over the pins. If the gage slides over any of the pins, replace the External Power Receptacle. These are the tasks:

External Power Receptacle Removal, TASK 24-41-11-000-801

External Power Receptacle Installation, TASK 24-41-11-400-801.

SUBTASK 24-41-11-410-005

(6) Close this access panel:

Number Name/Location

114AR External Power Receptacle Door

Lesson 11 Fire Protection

Section 5 Aviation Application: Read the following PP material with what you have learned.

13 PP 译文

14《故障隔离手册》（微课）

ENGINE AND APU FIRE CONTROL PANEL - REMOVAL/INSTALLATION

1. General

A. This procedure has these tasks:

(1) A Removal of the Engine and APU Fire Control Panel, Referred to As the Fire Control Panel.

(2) An Installation of the Engine and APU Fire Control Panel.

(3) A Removal of the Engine and APU Fire Shutoff Switch Assembly.

(4) An Installation of the Engine and APU Fire Shut off Switch Assembly.

B. The Engine and APU Fire Control Panel is installed on the P8 aisle control stand in the flight compartment.

TASK 26-00-01-000-801

2. Engine and APU Fire Control Panel Removal (Figure 401)

A. Location Zones.

Zone	Area
211	Flight Compartment - Left
212	Flight Compartment - Right

B. Prepare for the Removal of Engine and APU Fire Control Panel.

SUBTASK 26-00-01-860-001

(1) Open these circuit breakers and install safety tags:

a. F/O Electrical System Panel, P6-2.

Row	Col	Number	Name
A	19	C00388	FIRE PROTECTION DET OVHT WW WING BODY
A	21	C00396	FIRE PROT DETECTION MA WRN & CONT
A	22	C00407	FIRE PROTECTION DETECTION ENG 2
A	23	C00403	FIRE PROTECTION DETECTION APU
A	24	C00405	FIRE PROTECTION DETECTION ENG 1
B	19	C01344	APU FIRE SW POWER
B	20	C00297	FIRE PROTECTION EXTINGUISHERS RIGHT
B	21	C00452	FIRE PROTECTION EXTINGUISHERS APU
B	22	C00296	FIRE PROTECTION EXTINGUISHERS LEFT
B	23	C01022	FIRE PROTECTION EXTINGUISHERS ALTN R
B	24	C01021	FIRE PROTECTION EXTINGUISHERS ALTN L

b. F/O Electrical System Panel, P6-3.

Row	Col	Number	Name
B	3	C00360	FUEL SPAR VALVE ENG 2
B	4	C00359	FUEL SPAR VALVE ENG 1

SUBTASK 26-00-01-860-004

(2) If it is necessary, set the engine fuel levers to IDLE to get access to remove the Engine and APU Fire Control Panel.

C. Engine and APU Fire Control Panel Removal.

SUBTASK 26-00-01-020-001

(1) Loosen the four quarter turn fasteners securing the Engine and APU Fire Control Panel [1] to the frame.

SUBTASK 26-00-01-020-002

(2) Slide the Engine and APU Fire Control Panel [1] from the frame.

SUBTASK 26-00-01-020-003

(3) Tag and remove the five electrical connectors.

SUBTASK 26-00-01-860-005

(4) Make sure the engine fuel levers switch to CUTOFF.

Row	Col	Number	Name
A	19	C00388	FIRE PROTECTION DET OVHT WW WING BODY
A	21	C00396	FIRE PROT DETECTION MA WRN & CONT
A	22	C00407	FIRE PROTECTION DETECTION ENG 2
A	23	C00403	FIRE PROTECTION DETECTION APU
A	24	C00405	FIRE PROTECTION DETECTION ENG 1
B	19	C01344	APU FIRE SW POWER
B	20	C00297	FIRE PROTECTION EXTINGUISHERS RIGHT
B	21	C00452	FIRE PROTECTION EXTINGUISHERS APU
B	22	C00296	FIRE PROTECTION EXTINGUISHERS LEFT
B	23	C01022	FIRE PROTECTION EXTINGUISHERS ALTN RIGHT
B	24	C01021	FIRE PROTECTION EXTINGUISHERS ALTN LEFT
B	3	C00360	FUEL SPAR VALVE ENG 2
B	4	C00359	FUEL SPAR VALVE ENG 1

LAVATORY WASTE COMPARTMENT FIRE EXTINGUISHING BOTTLE INSPECTION/ CHECK

TASK 26-24-01-200-801 (Figure 201)

A. General.

This procedure is a scheduled maintenance task.

B. Location Zones.

Zone   Area

200   Upper Half of Fuselage

C. Procedure.

SUBTASK 26-24-01-280-002

(1) Open the waste compartment door.

SUBTASK 26-24-01-280-006

(2) Open the waste container door to get access to the fire extinguishing bottle on the inside of the waste compartment.

SUBTASK 26-24-01-020-006

(3) Remove the waste container.

SUBTASK 26-24-01-960-001

(4) Examine the fusible tips on the discharge tubes.

Replace the fire extinguishing bottle and the temperature indicator strip if the fusible tips are melted.

NOTE: The temperature indicator strip is located above the waste container.

NOTE: The temperature indicator strip is located on the front of the waste container door.

SUBTASK 26-24-01-960-002

(5) Examine the temperature indicator strip to see if one or more of the temperature indicators changed from white to black.

If one or more of the temperature indicator slips changed from white to black, replace the temperature indicator strip.

SUBTASK 26-24-01-280-004

(6) Examine the fire extinguishing bottle for corrosion, scratches, or dents.

Replace the fire extinguishing bottle if any dents are deeper than 1/16 in (2 mm)per 1 in (25 mm) of average dent diameter or if any scratches are deeper than 0.004 in (0.102 mm).

SUBTASK 26-24-01-020-008

(7) Install the waste container in the waste compartment.

SUBTASK 26-24-01-280-007

(8) Close the waste container door.

SUBTASK 26-24-01-280-005

(9) Close the waste compartment door.

Lesson 12 Lighting System

Section 5 Aviation Application: Read the following PP material with what you have learned.

13 PP 译文   14 飞机的种类
                      （微课）

MASTER WARNING AND CAUTION SYSTEM ANNUNCIATOR - LAMP REPLACEMENT

TASK 33-11-00-960-804

1. Master Warning And Caution System Annunciator - Lamp Replacement Figure 204

A. References

Reference	Title
24-22-00-860-811	Supply Electrical Power (P/B 201)
24-22-00-860-812	Remove Electrical Power (P/B 201)

B. Expendables/Parts.

AMM Item	Description	AIPC Reference	AIPC Effectivity
21	Lamp	33-14-00-02-035	GUN ALL
		33-14-00-03-105	GUN 008-030, 101-121
		33-14-00-03-245	GUN 031-099, 122-131
		33-14-00-03A-100	GUN 001-007, 155-157,201-999
		33-14-00-09-105	GUN 008-099, 101-131
		33-14-00-09-130	GUN 008-099, 101-131
		33-14-00-09A-105	GUN 001-007, 155-157,201-999
		33-14-00-09A-150	GUN 001-007, 155-157, 201-999
		33-14-00-20-010	GUN 001-007, 155-157, 201-999

C. Location Zones.

Zone	Area
211	Flight Compartment - Left
212	Flight Compartment - Right

D. Procedure.

SUBTASK 33-11-00-860-013

(1) Open these circuit breakers and install safety tags: F/O Electrical System Panel, P6-3

Row	Col	Number	Name
B	12	C00132	MASTER CAUTION ANNUNCIATOR BUS 1
B	13	C00131	MASTER CAUTION ANNUNCIATOR BAT

SUBTASK 33-11-00-960-004

(2) Do these steps to replace the lamp [21]:

CAUTION: REMOVE THE LAMP HOUSING WITH YOUR FINGERS ONLY. IF YOU

USE A TOOL TO HOLD THE LAMP HOUSING, THE TOOL CAN CAUSE DAMAGE TO THE HOUSING.

    a. Pull the lamp housing [23] from the annunciator assembly [22].

    b. Remove the defective lamp [21].

    c. Put the new lamp [21] into the lamp housing [23].

    d. Put the lamp housing [23] into the annunciator assembly [22].

    e. Make sure that the lamp housing [23] engages correctly in the annunciator assembly [22].

SUBTASK 33-11-00-860-014

(3) Do this task: Supply Electrical Power, TASK 24-22-00-860-811.

SUBTASK 33-11-00-860-015

(4) Remove the safety tag and close the applicable circuit breaker: F/O Electrical System Panel, P6-3.

Row	Col	Number	Name
B	12	C00132	MASTER CAUTION ANNUNCIATOR BUS 1
B	13	C00131	MASTER CAUTION ANNUNCIATOR BAT

SUBTASK 33-11-00-710-004

(5) Do a test of the new lamp [21]:

    a. At the captain's main instrument panel, P1, set the switch for the master dim and test lights to TEST.

    b. Make sure that the replaced lamp [21] is on.

    c. Set the switch to the off position.

SUBTASK 33-11-00-860-016

(6) Do this task: Remove Electrical Power, TASK 24-22-00-860-812.

FLOOR PROXIMITY LIGHT - LIGHT ASSEMBLY REPLACEMENT

TASK 33-51-14-960-802 (Figure 202)

A. References.

Reference	Title
WDM 33-51-21	Wiring Diagram Manual

B. Location Zones.

Zone	Area
200	Upper Half of Fuselage

C. Procedure.

SUBTASK 33-51-14-960-002

(1) Do these steps to replace the applicable light assembly (WDM 33-51-21):

a. Remove the fasteners, [21] or [24].

b. Pull out the light assembly, [22] or [25].

c. Disconnect the electrical wires, [23] or [26].

d. Connect the electrical wires, [23] or [26] to the new light assembly.

Make sure the new light assembly includes a lamp (TASK 33-51-14-960-801).

e. Put in the new light assembly, [22] or [25].

Make sure the electrical wires are installed securely in their usual position.

f. Install the fasteners, [21] or [24].

SUBTASK 33-51-14-710-002

(2) Do a test of the new light assembly:

NOTE: The emergency lights are energized by the power supplies. Use the lights for a minimum time to do the test.

a. At the attendant panel in the passenger compartment, set the emergency light switch to the on mode.

Make sure the light comes on.

b. Set the switch to the off mode.

Make sure all emergency lights and signs go off.

Lesson 13　Navigation System

Section 5　Aviation Application: Read the following PP material with what you have learned.

13 PP 译文

14 波音 737 经典的
一代 VS 次世代
（微课）

PITOT SYSTEM - DETAILED INSPECTION OF DRAINS

TASK 34-11-00-210-801

A. Location Zones.

Zone　Area

112　　Area Forward of Nose Landing Gear Wheel Well

121　　Forward Cargo Compartment - Left

122　　Forward Cargo Compartment - Right

211　　Flight Compartment - Left

212　　Flight Compartment - Right

231　　Forward Passenger Compartment - Forward Entry Door to Sta 663.75- Left

232　　Forward Passenger Compartment - Forward Entry Door to Sta 663.75- Right

B. Procedure.

SUBTASK 34-11-00-210-001

(1) Do a detailed inspection for moisture for the Captains and First Officers pitot system drains.

NOTE: The alternate pitot system does not have a drain fitting. The probe is at the lowest part of the system line so that moisture can drain from the probe.

SUBTASK 34-11-00-680-001

(2) If you find moisture in at least one of the locations above, do this task: Static and Total Air Pressure System - Servicing, TASK 34-11-00-170-801.

FMCS CONTROL DISPLAY UNIT - SERVICING

1. General

A. This procedure contains the task for the CDU lamp test.

B. You can find the FMCS control display units (CDUs) in the flight compartment on the forward electronic panel, P9. The CDU has four annunciators (DSPY or CALL, FAIL, MSG, and OFST) and a lighted EXEC key. Two of the annunciators are on the left of the CDU (DSPY or CALL/FAIL), and two are on the right of the CDU(MSG/OFST).

C. The CDU has lamps for the annunciators and the EXEC key that are not line replaceable. This requires replacement of the keyboard or the CDU.

TASK 34-61-01-710-801

2. CDU Lamp Test

A. References.

Reference   Title

24-22-00-860-811 Supply Electrical Power (P/B 201)

24-22-00-860-812 Remove Electrical Power (P/B 201)

B. Location Zones.

Zone   Area

211   Flight Compartment - Left

212   Flight Compartment - Right

C. Procedure.

SUBTASK 34-61-01-860-008

(1) Do this task: Supply Electrical Power, TASK 24-22-00-860-811.

SUBTASK 34-61-01-710-002

(2) Do the CDU Lamp Test:

a. Hold the LIGHTS switch in the TEST position. You can find the LIGHTS switch on the pilots' center instrument panel (P2).

b. Make sure that the DSPY (or CALL), FAIL, MSG, OFST, and EXEC lights on the two CDUs come on.

c. Set the LIGHTS switch to the BRT or DIM position if it is necessary.

D. Put the Airplane Back to Its Usual Condition.

SUBTASK 34-61-01-860-018

Do this task: Remove Electrical Power, TASK 24-22-00-860-812

Lesson 14 Warning System

Section 5 Aviation Application: Read the following PP material with what you have learned.

13 PP 译文 14 旋翼飞机主要
结构（微课）

APU FIRE WARNING HORN - REMOVAL/INSTALLATION

1. General

A. This procedure has these tasks:

(1) The Removal of the APU Fire Warning Horn

(2) The Installation of the APU Fire Warning Horn

B. The APU Fire Warning Horn M277 is in the P28 APU Fire Control Panel.

TASK 26-15-02-000-801

2. APU Fire Warning Horn Removal

A. Location Zones.

Zone Area

134 Main Landing Gear Wheel Well, Body Station 663.75 to Body Station 727.00 - Right

B. APU Fire Warning Horn Removal.

SUBTASK 26-15-02-010-001

(1) Open this circuit breaker and install safety tag: F/O Electrical System Panel, P6-2.

Row   Col   Number   Name

A   23   C00403   FIRE PROTECTION DETECTION APU

SUBTASK 26-15-02-020-001

(2) Tag each electrical wire with its location.

SUBTASK 26-15-02-020-002

(3) Loosen terminal lugs.

SUBTASK 26-15-02-020-003

(4) Disconnect the electrical wires from the APU Fire Warning Horn.

SUBTASK 26-15-02-020-004

(5) Remove four screws and four washers that connect the horn to the P28 panel.

SUBTASK 26-15-02-020-005

(6) Remove APU Fire Warning Horn.

STALL WARNING SYSTEM - OPERATIONAL TEST

TASK 27-32-00-710-801

A. General.

This procedure is a scheduled maintenance task.

B. References.

Reference   Title

24-22-00-860-813   Supply External Power (P/B 201)

24-22-00-860-814   Remove External Power (P/B 201)

C. Location Zones.

Zone   Area

212   Flight Compartment – Right

D. Procedure.

SUBTASK 27-32-00-860-001

(1) Do this task: Supply External Power, TASK 24-22-00-860-813

SUBTASK 27-32-00-710-001

(2) Do a test of the number 1 stall warning system:

NOTE: The amber SPD LIM light on the on-side Primary Flight Display will come on. The MASTER CAUTION light will come on if the Yaw Damper switch is "disengaged/off". This will not occur on 285A1010-7 and 285A1010-9 SMYD units. A SPD LIM annunciation that stays on more than 5 s after the stick shaker operates could be a problem.

a. Push and hold the STALL WARNING TEST NO. 1 switch on the stall warning test module (P5-18 panel).

b. Make sure that the captain's control column shaker operates.

c. Release the switch.

d. Make sure that the shaker stops.

SUBTASK 27-32-00-710-002

(3) Do a test of the number 2 stall warning system:

NOTE: The amber SPD LIM light on the on-side Primary Flight Display will come on. The

MASTER CAUTION light will come on if the Yaw Damper switch is "disengaged/off". This will not occur on 285A1010-7 and 285A1010-9 SMYD units. A SPD LIM annunciation that stays on more than 5 s after the stick shaker operates could be a problem.

a. Push and hold the STALL WARNING TEST NO. 2 switch on the stall warning test module(P5-18 panel).

b. Make sure that the first officer's control column shaker operates.

c. Release the switch.

d. Make sure that the shaker stops.

SUBTASK 27-32-00-860-002

(4) If electrical power is no longer needed: do this task: Remove External Power.

Appendix IV   My Leaflet Workbook

Lesson 1   Air Conditioning and Pressurization System

 Learning Reflection

【Learning Achievement】
1. The terms I have memorized:

为建设航空强国而
努力奋斗

2.The sentences I have understood:

【Learning to be Improved】
1. The terms I can't memorize:

2. The sentences I don't understand:

Lesson 2　Flight Control System

 Learning Reflection

【Learning Achievement】

1. The terms I have memorized:

从民航大国向民航
强国迈进

2. The sentences I have understood:

【Learning to be Improved】

1. The terms I can't memorize:

2. The sentences I don't understand:

Lesson 3   Fuel System

【Learning Achievement】

1. The terms I have memorized:

强国有我成为
2021 年度十大
网络用语之一

2. The sentences I have understood:

【Learning to be Improved】

1. The terms I can't memorize:

2. The sentences I don't understand:

Lesson 4 Hydraulic System

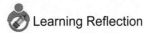 Learning Reflection

【Learning Achievement】

1. The terms I have memorized:

中国航天成为
2021 年度十大
流行语之一

2. The sentences I have understood:

【Learning to be Improved】

1. The terms I can't memorize:

2. The sentences I don't understand:

Lesson 5   Pneumatic System

Learning Reflection

【 Learning Achievement 】

1. The terms I have memorized:

传承罗阳精神，建
设航空强国

2. The sentences I have understood:

【 Learning to be Improved 】

1. The terms I can't memorize:

2. The sentences I don't understand:

Lesson 6   Water and Waste System

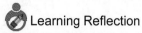Learning Reflection

【Learning Achievement】

1. The terms I have memorized:

匠心筑梦　匠艺强国

2. The sentences I have understood:

【Learning to be Improved】

1. The terms I can't memorize:

2. The sentences I don't understand:

Lesson 7　Engine Oil

【Learning Achievement 】

1. The terms I have memorized:

国产大飞机订单超
1000，后发企业
怎么打侧翼战

2. The sentences I have understood:

【Learning to be Improved 】

1. The terms I can't memorize:

2. The sentences I don't understand:

Lesson 8　Communications System

 Learning Reflection

【Learning Achievement 】

1. The terms I have memorized:

国产大飞机 C919
飞行员已就位，旅
客最快何时可乘坐

2. The sentences I have understood:

【Learning to be Improved 】

1. The terms I can't memorize:

2. The sentences I don't understand:

Lesson 9   Electrical Power System

Learning Reflection

【 Learning Achievement 】

1. The terms I have memorized:

国产大飞机 C919
完成取证试飞

2. The sentences I have understood:

【 Learning to be Improved 】

1. The terms I can't memorize:

2. The sentences I don't understand:

Lesson 10　External Power

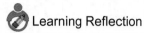

Learning Reflection

【Learning Achievement 】

1. The terms I have memorized:

秉承航空强国使
命，全面推进高
质量发展

2. The sentences I have understood:

【Learning to be Improved 】

1. The terms I can't memorize:

2. The sentences I don't understand:

Lesson 11   Fire Protection

Learning Reflection

【Learning Achievement】
1. The terms I have memorized:

航空强国　志翼起
向未来

2. The sentences I have understood:

【Learning to be Improved】
1. The terms I can't memorize:

2. The sentences I don't understand:

Lesson 12　Lighting System

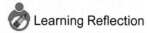 Learning Reflection

【Learning Achievement】

1. The terms I have memorized:

航空工业沈飞刘洪
波：披荆斩棘勇担
当的飞机修理人

2. The sentences I have understood:

【Learning to be Improved】

1. The terms I can't memorize:

2. The sentences I don't understand:

Lesson 13　Navigation System

【 Learning Achievement 】

1. The terms I have memorized:

航空报国　航空强国

2. The sentences I have understood:

【 Learning to be Improved 】

1. The terms I can't memorize:

2. The sentences I don't understand:

Lesson 14　Warning System

 Learning Reflection

【Learning Achievement】

1. The terms I have memorized:

长沙航院学子桑云逸

2. The sentences I have understood:

【Learning to be Improved】

1. The terms I can't memorize:

2. The sentences I don't understand:

References

[1] 白杰，张帆. 民航机务英语教程 [M]. 北京：中国民航出版社，1997.

[2] 蒋陵平. 燃气涡轮发动机（ME-TA,TH）[M]. 2 版. 北京：清华大学出版社，2016.

[3] 张铁纯. 涡轮发动机飞机结构与系统(ME-TA)(上)[M]. 2 版. 北京:清华大学出版社，2017.

[4] 任仁良. 涡轮发动机飞机结构与系统(ME-TA)(下)[M]. 2 版. 北京:清华大学出版社，2017.

[5] 任仁良. 涡轮发动机飞机结构与系统（AV）（下）[M]. 北京：清华大学出版社，2017.

[6] 张鹏. 涡轮发动机飞机结构与系统（AV）（上）[M]. 北京：清华大学出版社，2017.

[7] Boeing Editors. AMM[Z]. Boeing Company, 2001.

[8] TOOLEY M, WYATTD D. Aircraft Communications and Navigation Systems: Principles, Operation and Maintenance[M]. Routledge Taylor & Francis Group, 2007.

[9] CRANE D. Airframe Volume 2: Systems [M]. 3r ed. Aviation Supplies &Academics Inc, 2007.

[10] EISMIN T K. Aircraft Electricity and Electronics[M]. Tata McGraw-Hill Education, 2014.

[11] DINGLE L, TOOLEY M. Aircraft Engingeering Principles[M]. Butterworth Heinemann Linacre House, 2004.

读书笔记